Making Miniatures With 3D Printers

Designing for dollhouse, book nooks and dioramas

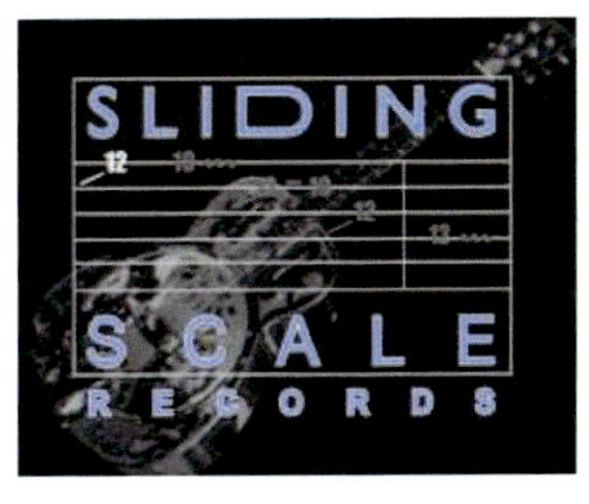

slidingscale.co.uk

Dedication

To my lovely wife Angie who insisted I write a book.

Publishing Data

First edition published 2024 (SSFF01)

Text copyright Frank Fisher and Angie Scarr

Illustration, composites and photographs copyright Frank Fisher

Design by Frank Fisher and Angie Scarr

Plaza De Andalucía 1, Campofrío, 21668, Huelva, Spain.

ISBN 978-84-126023-7-1

Contents

Foreword by Angie Scarr

If you're anything at all like me your first thought might be "That's tech...I can't do that".

Fortunately I have always had Frank to help me when I need to turn design ideas into reality.

But what about people who don't have a tame computer geek?

So I said to Frank "Hey, what about writing a book as if you were writing it for me. So that people like me, who really want to come up with their own designs for 3D printers can cut through all the scary technical stuff and learn to do it bit-by-bit?"

So that is where this book came from. It's step by step and pretty much in our 'house style' and I hope it will lead you by the hand through the technical stuff that scares me and might intimidate you. It's written for people like us!

I haven't had to do everything in this book to prove it can be done, but I did do the plant pot. How exciting! Maybe I could do it after all! If you have never designed for a 3D printer before, we both recommend that you do the plant pot yourself before trying anything else. Actually even if you have, doing the plant pot will help you get into Frank's style of teaching.

This book should help you to imagine how you can create your own files and of course adapt bought files.

'Back to Front'

Frank decided to put the fun designing bit first, and then added all the technical stuff at the back so you can use it as a reference book too, and dip into that whenever you need to learn more about how to use the software and about your machine, its capabilities and its weaknesses.

I hope whatever level you want to reach, this book helps you take your new printer out of the box if you have a new one. Or, if not, for the price of a book it helps you to decide whether you will be able to understand how to use and choose.

In fact, you don't even need to buy a printer if you have a friend who is happy to test your designs! You can even use a service to print for you and post the results.

So, throw yourself into the designing and when you are ready to print you will be more ready.

Be prepared for frustration at machine failures, and piles of wasted filament. We all have them!

Frank is very candid throughout the book about his errors and tech fails. Everyone has them and our 'house style' is not to pretend that everything works fine every time. It doesn't. Being an expert comes from making a lot of mistakes!

Frank recently bypassed the worst of this by getting his second machine rigged up with a camera to record progress and a remote control to instantly stop a failing print. Of course at one point in the writing of this book the second machine failed (being fixed now) and he had to go back to the old machine. You will follow him through his tech frustrations.

3D printing is still a relatively new hobby, and machines sometimes need a manufacturer fix too, so getting to know your machine and joining groups of other interested makers is always a good idea.

I hope this book gives you the confidence and the immense satisfaction to be had from knowing that at least part of your design is all yours!

Introduction by Frank Fisher

The book is set out assuming you have unboxed the printer, tested it and got as far as a successful test print. If you haven't set up the printer then go to the end of the book to work through all that - there's also quick start sections on making models and drawing SVGs. If you are set up and ready then let's make something right now and print it … we will start with projects you can finish in minutes and lead up to the larger and more sophisticated concepts.

- Make a plant pot from cones and a disc
- Make a square planter in a similar way
- Make boxes and crates with a fold
- Draw a metalwork bracket for a hanging basket
- Make a shop table with drawers
- Make a shop front with columns, window bay and opening door.
- Make a shop inner with shelving, display cabinet and cash register
- Make vases and decorative items

And more …

Designed for miniature

There are a lot of free STLs available for full size objects that you can scale down, but at a certain scale you may find these models break down. They may also be from a different style or period to what you need for a project. At this point it is often easier to start from scratch working at the scale you need and taking into account the limitations of your device. For example if you have a 0.4mm filament nozzle fitted nothing will be smaller than that so it isn't worth designing finer unless you intend to change it. It is also best to set the slicer to round up to that thickness rather than let the walls of a design fail. You can still step between layers at a finer resolution

e.g. 0.1mm, but the width of each line will be 0.4mm with a standard nozle.

Go right back to basics ...

and don't design without a brief.

Before making pieces I like to have the entire idea sketched out at least. It's even better if you do this at scale in a vector program such as Inkscape, as you can take pieces of this sketch out and use them directly in the design.

For the shop I had a front designed already which gave me the rough dimensions for the inside - the outside dimensions minus the depth of the side walls.

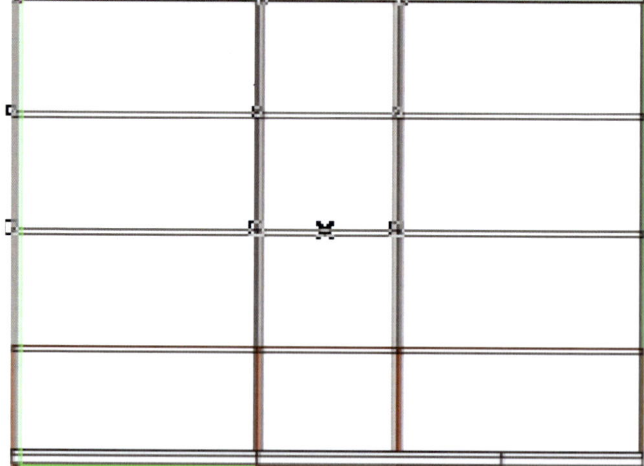

Angie wanted full wall to ceiling shelving running right across the back of the shop or the side of a book nook. Also it was to be modular so sections such as smaller drawer shelves or ribbon holders could be moved around inside the framework to suit the final design.

The bottom shelf is best sited at counter height to match other units such as the table and display unit.

Book nooks and dioramas

I am giving an example of a miniature shop in 12th scale but you can recombine and redesign the parts for other sizes. As an example I took the 12th scale parts and scaled them to 75% which let me take the door and one window frame to make a nook. Ideally if you are starting the design to this end you would cut the width of the window frame and deepen the room too so I took one pane of glass out and extended the walls to full bookshelf depth.

Nook example design 25cm/9.5" deep,15cm/6" wide and 21cm/8" high.

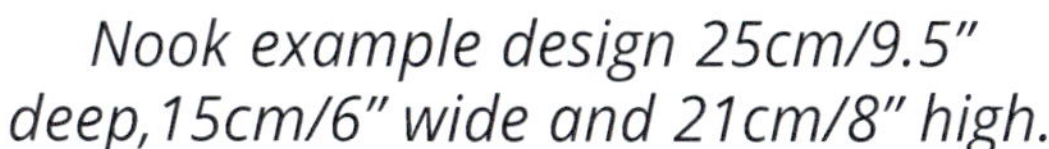

Part 1 - simple projects

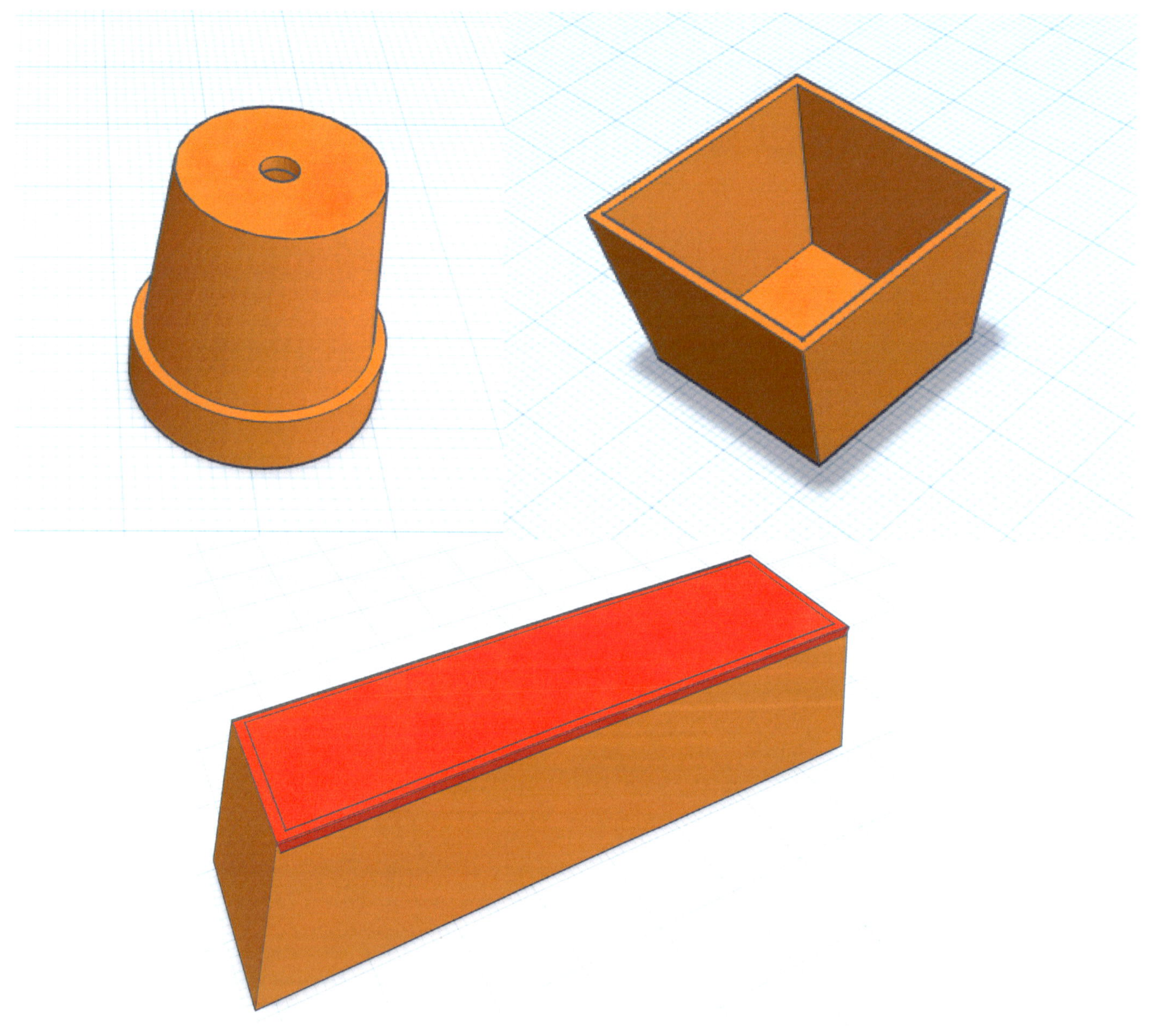

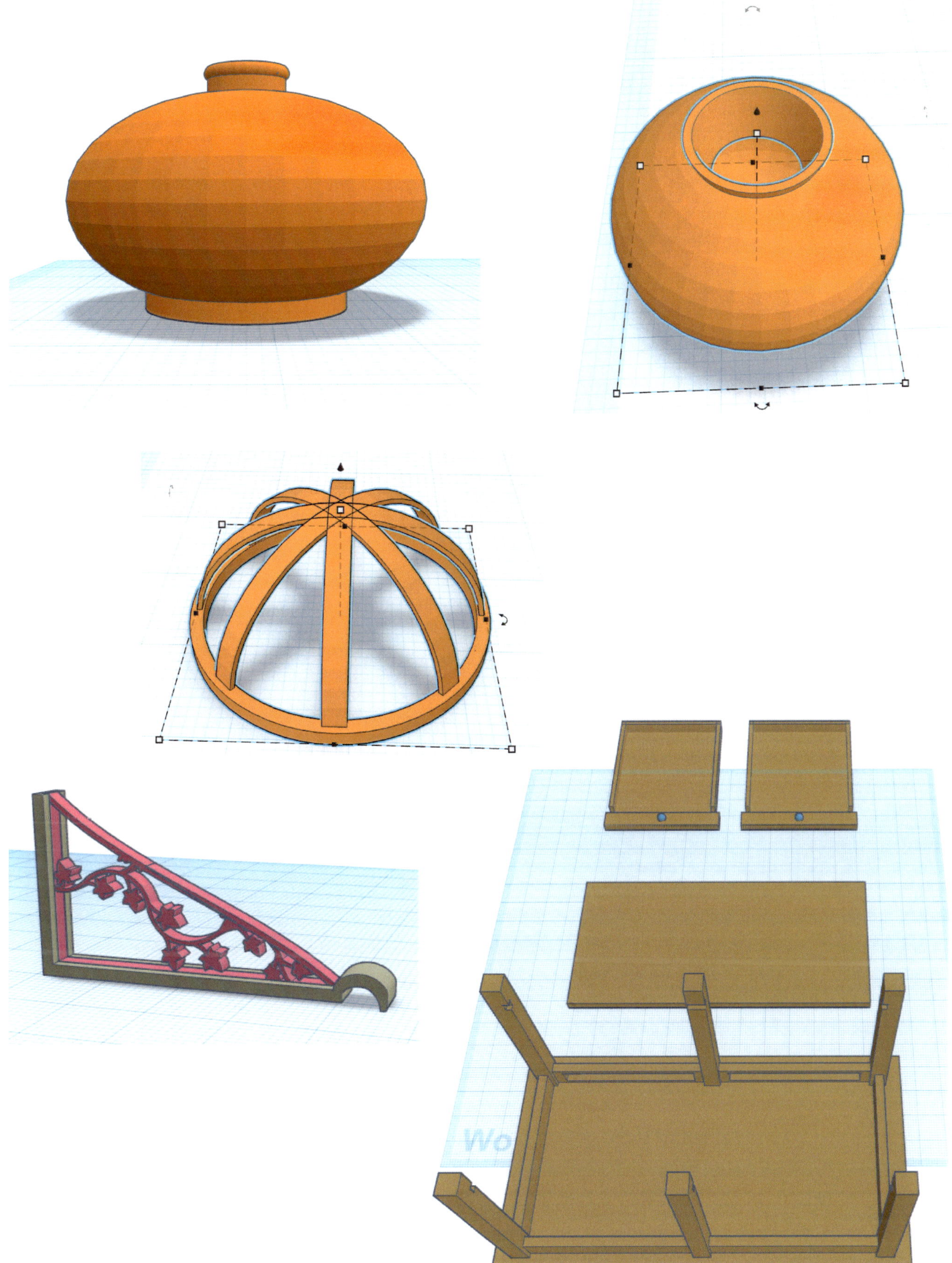

Plant pot

I'm going to take you through a really basic project in the free software Tinkercad, though any other 3D software is fine too.

This is to give the absolute beginners the confidence to find similar functions on whichever software they choose to use. In part 4 you will find more detailed info on how to use Tinkercad as well as Inkscape and Prusaslicer - three popular software programs you can use for free.

Create a free account on Tinkercad first and then we can start.

Tinkercad helps you to make objects from basic shapes like cubes, cones and spheres, which are called "primitives".

If you're an absolute beginner make sure you follow the instructions below step by step and line by line. If you make a mistake you can press undo until you're back to something you're sure about.

If you're a little more experienced you can just skip lightly through it but we still recommend you do this project just to help you get into our way of doing things.

We won't be repeating this level of information on where to find various functions except in part 4 where you'll find all these basics which you can refer to at any time.

The first thing you will need to do if you are confident is to decide the height and diameter of the pot. Otherwise follow our suggestions exactly.

First, the purple cone will be the main body of the pot. Drag a purple cone into the middle of your workspace.

Make it 19mm diameter. You need to click on one of the white squares around the base and when you click it the number box will appear. You can either drag these squares to change

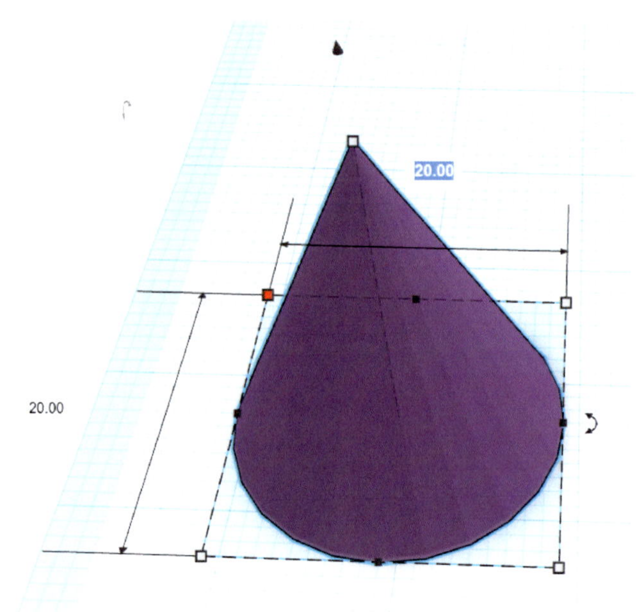

the size, or simply type the numbers you want into the number boxes. Make it 95mm high using the centre white square (on the top). You might have to use the Fit All button 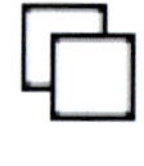

Next copy the cone by using the copy button on the top left hand corner of the screen. Then Paste. Shrink the copy to 17mm diameter and 93mm high and 

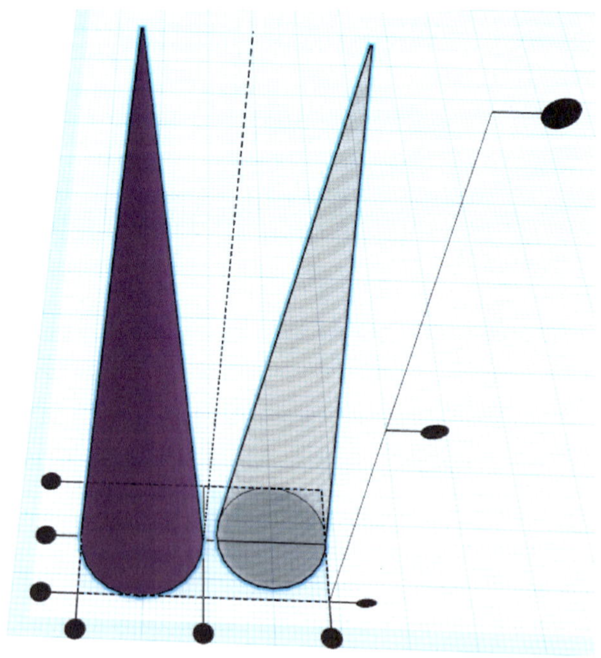

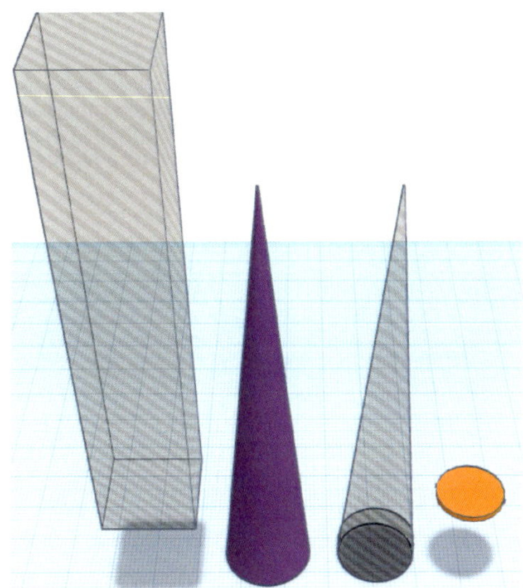

then turn it into a 'hole'. You'll find that button on the top right of your screen.

Now automatically centre the two cones together by selecting both. To do this sweep across both with the button held down and let go when you have a line or box touching both. Then go to the top right hand side above the hole button and you'll see an align button which looks like this.

One of the easiest things to confuse is the centre edge squares with the axis alignment circles.

Click on the two axes you want to centre in this case its middle front (X) and middle side (Y). The hole will sit inside the cone invisibly for now. It will leave a 1mm thick wall. All you can see is one purple cone, but it does have a hole inside!

Click group to keep the two together.

Now drag an orange cylinder and pull it down to 1mm thick (height) and 15mm in diameter. Then raise it to a height of 19mm from the base using the up arrow which starts black but turns red when you drag it. You should see a shadow underneath. That's just to remind you what you've done!

Drag a hole cube over to your workspace and make it 20mm x 20mm x 80mm high. Raise it to a height of 19mm above the base. Bring this

all into view with Fit All in order to find the little up arrow.

If you now use the align tool again you can centre the cones and hole cube using the X and Y axes. (Don't touch the height Z) so that the box appears to be around, but a bit above the bottom of the purple cone.

Select both and use the group function again and you should see an upside down pot shape, with no base.

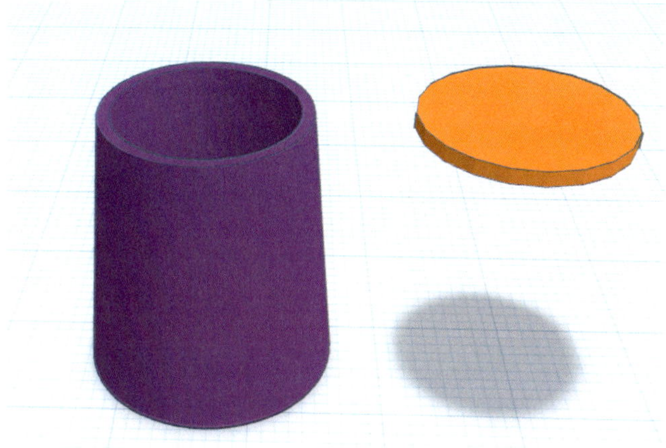

Now centre the orange disc with it and group again. This leaves a finished pot shape. Don't believe me? Take a look at the underneath using the VIEW cube which is at the top left and looks like this.

Go ahead. Play with looking at the shape from all angles. Drag the cube left to right, up and down. Magic, isn't it? If you've made it this far, you have climbed the worst of the tech fear mountains and have earned your novice 'techy' badge!

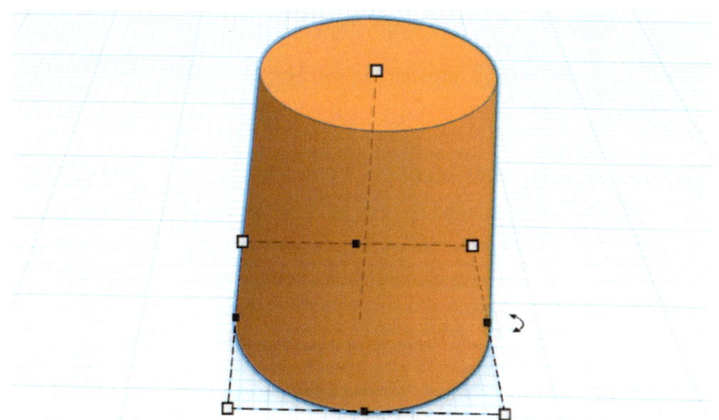

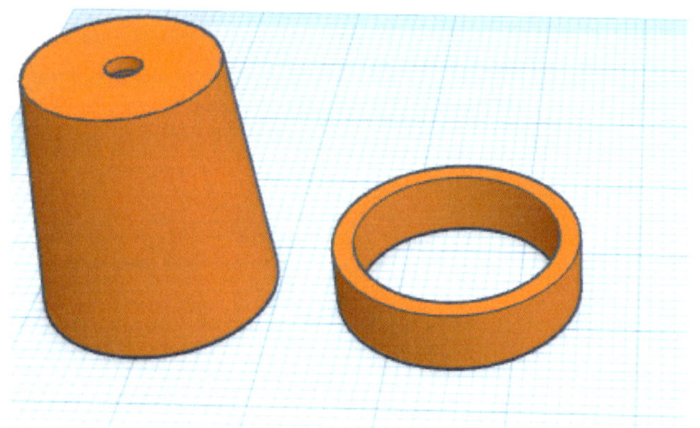

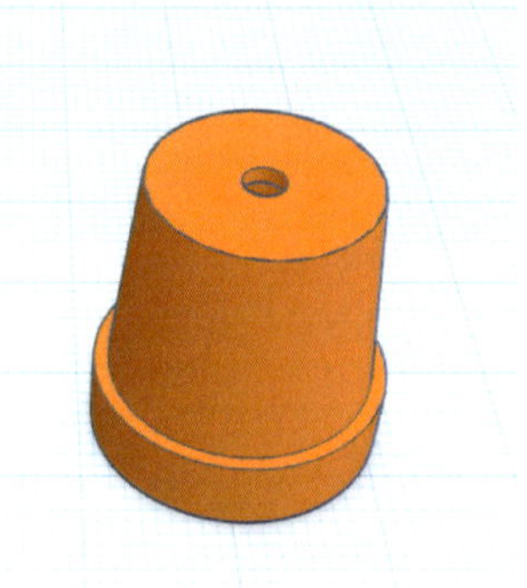

You can make it better by adding a small hole in the bottom using a cylinder centred and turned into a hole. Most printers are fine with this, if yours isn't then a craft drill will do the job afterwards.

Also I like to have a collar on the top of the pot to look like an old terracotta pot. Make a cylinder of diameter 21mm with a hole of diameter 19mm, and height of 5mm. Align and group this to the simple pot.

Slicing and printing the model

I am now going to start assuming you already have the software installed, setup and tested, if not see the sections later in the book on doing this.

Using Tinkercad as an example again, click on the pot then click Export from the top right of the screen.

Include->the selected shape. *This is a good default so you can have multiple pieces on the workboard at once - useful for getting dimensions and alignment perfect - but only slice the pieces you need.*

Click STL and give it a name.

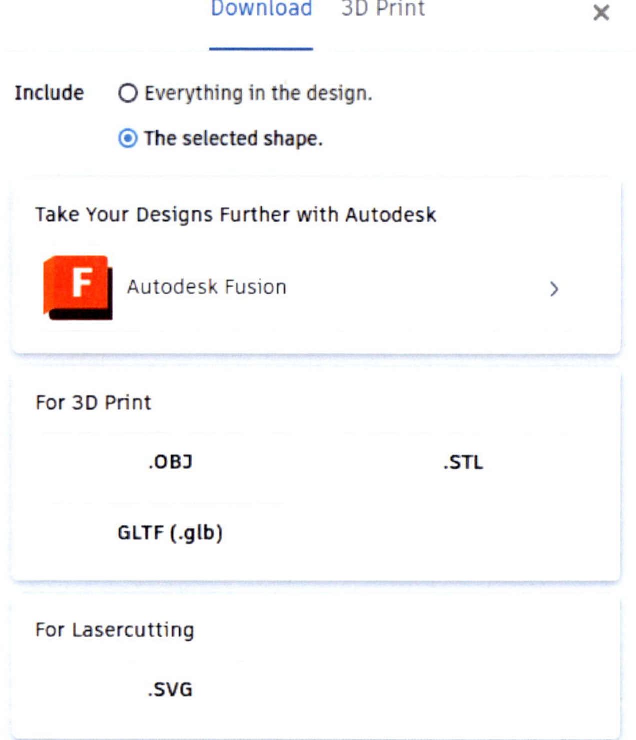

Extra equipment

- *Nail artist's files and buffers*
- *Small scissors*
- *Glue*
- *Baby oil or alcohol to restore shine*

Depending on your printer bed you may have to sand some or most of the edges of the objects you print.

If you needed to add Brim to print a tube on it's side, for instance, you will need to clear away the Brim and smooth it down.

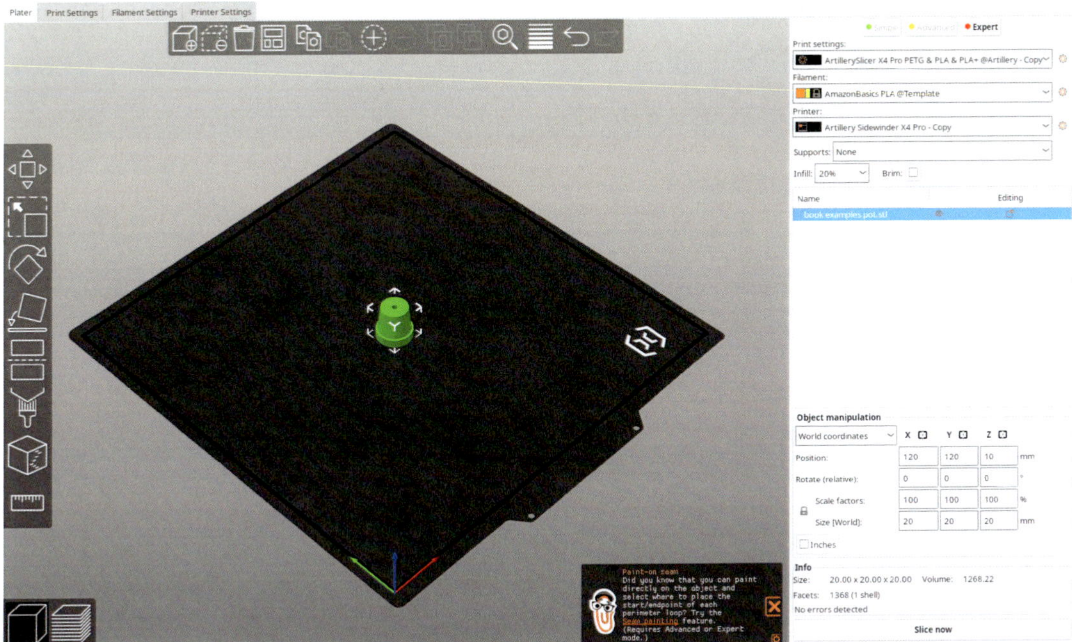

Go to your slicer, the free Prusa Slicer is used in this example. It should show an empty printing mat for your machine.

Go to file->Import and choose the STL file you just saved above. It should appear in the centre of the mat.

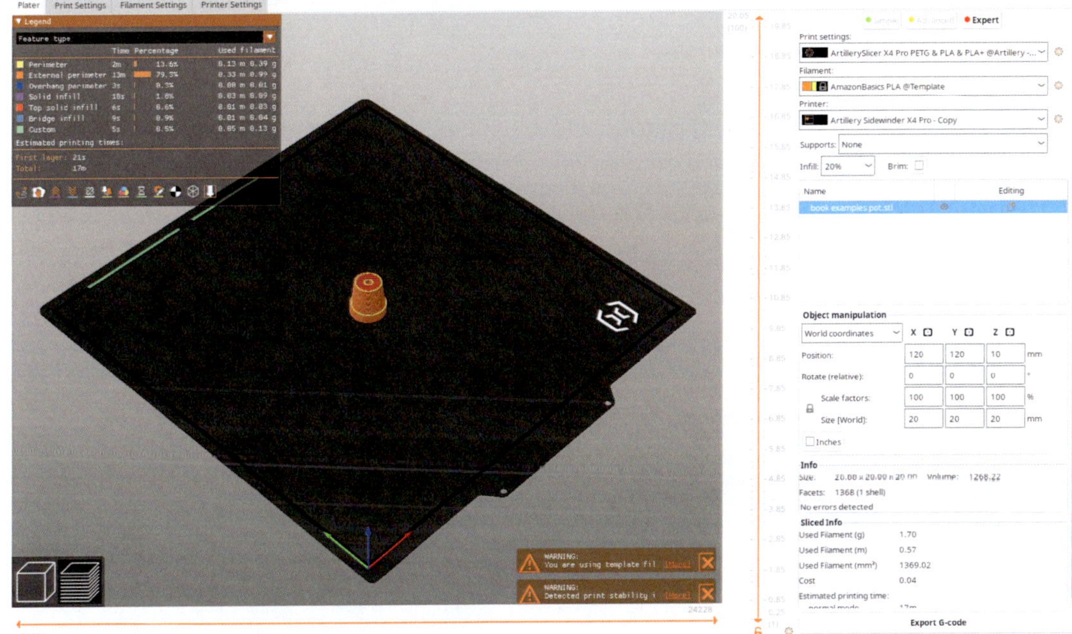

If the filament settings etc. are OK (see top right of the screen) then you can click Slice Now at the bottom right.

You should now have time and filament usage estimates for the model and the option to Export G-code, do this and choose a filename.

If you end up with a folder of G-code and can't remember what is what anymore there is a free G-code viewer from Prusa which will show you what the file makes.

More shapes

By using a pyramid of 30mm W x 30mm L x 100mm high for the base object you can make a square planter using the same principles.

Make a hole pyramid of 28mm W x 28mm L x 98mm high, and cut off using a hole cube of 32mm x 32mm x 80mm at height

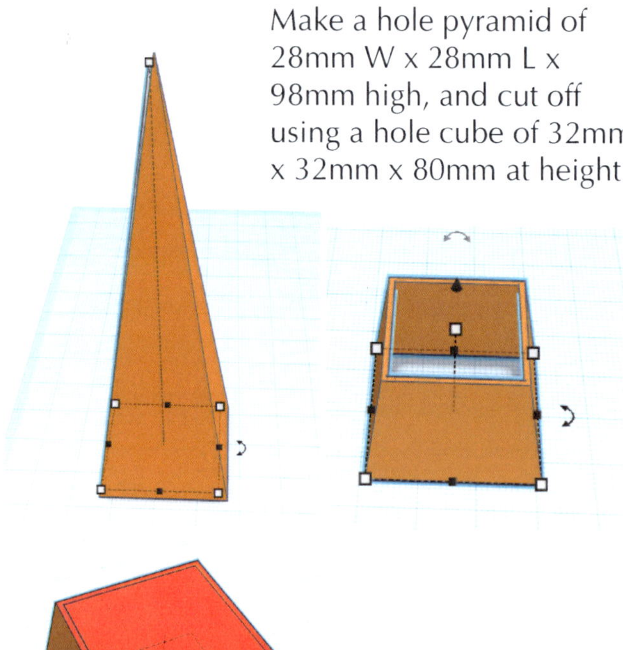

20mm. You can use the alignment tool to put the cube to the top then centre everything.

Make a cube of 24mm square and 1mm high. Raise it to 19mm, centre it to the sides and group.

Optionally add a drainage hole(s) to the design.

If you are not adding a collar you can flip the designs around 180 degrees to print upright.

NOTE: it is important to make sure items are aligned to the "ground" (Z=0) before slicing or lots of problems can happen.

Next steps

If you make the pyramid 20mm W x 80mm L x 100mm high then you have a window box.

The base will be a cube of 18mm x 64mm x 1mm.

You can make tall rounded or squared flower buckets and any number of other items from this quick principle. Simply decide your height, base diameter and top diameter and you can adjust the primitives until they fit, then crop them to shape with hole primitives.

Vases 1

Simple vase shapes

You can reproduce many of the shapes you want from a combination of primitives plus holes. Get a Front view of the workspace and place a small cylinder for the base, probably 2mm high and your choice of diameter; my example is 10.5mm.

Next add a sphere and pull its sides and height until you have a shape you like, I went for 20mm diameter and 14mm high.

For a simple rim I have put an upside down cone in the top for a couple of mm, 9mm diameter and 14mm high..

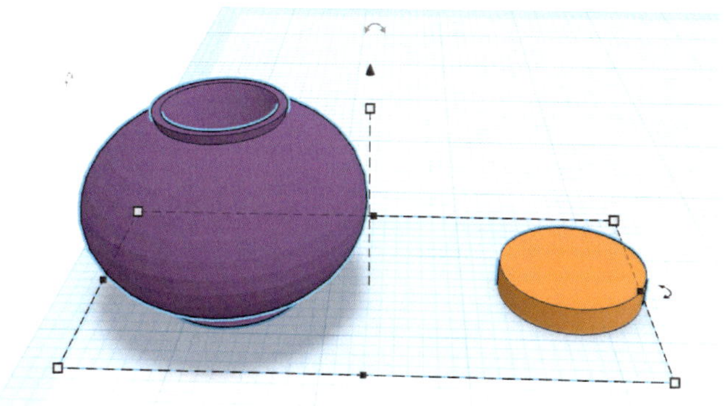

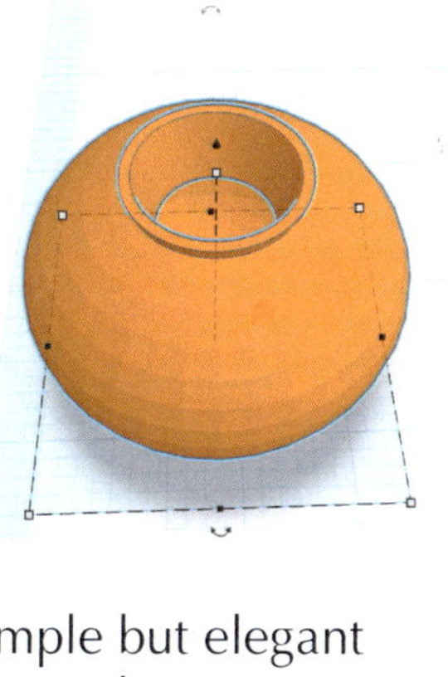

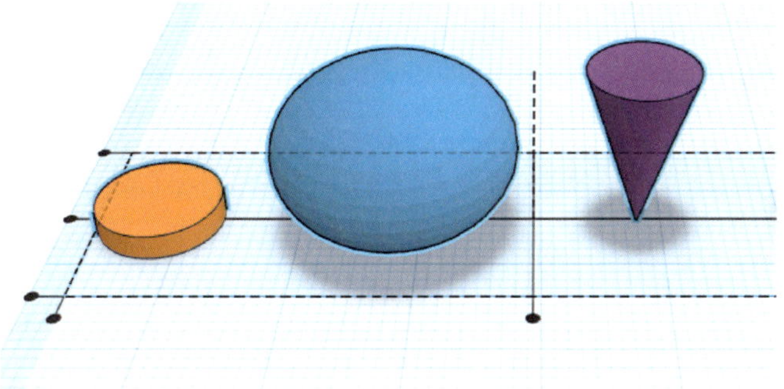

You should now have a simple but elegant shape. Export it to the slicer and use 100% infill. If this prints you are done, if not correct any errors until it is perfect.

You can now make copies of each shape except the base, make them at least 2mm smaller all around, more if the curves are extreme, and convert to holes. You will remember from the plant pot examples that this leaves you a 1mm wall all around the hollow shape. If you want it thinner you can but be aware at a certain size the print will break as it will not have enough wall to build the next layer on as it curves.

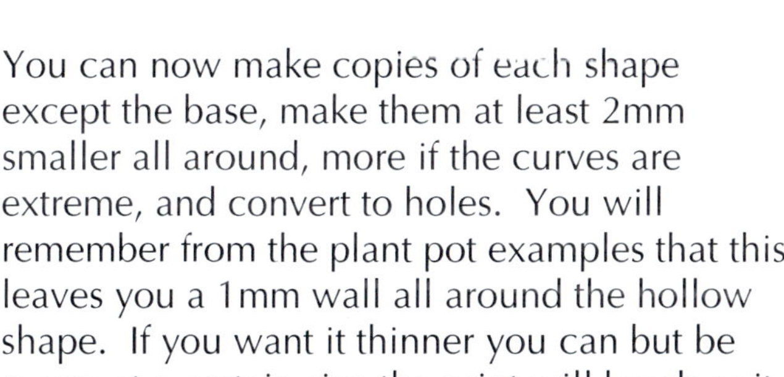

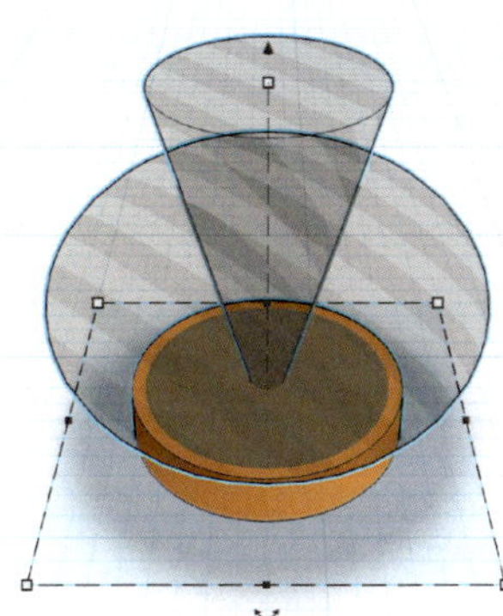

will not have enough wall to build the next layer on as it curves.

Align all except the base to the centre in X,Y and group. Now centre and group the base - this prevents holes in the base.

More shapes

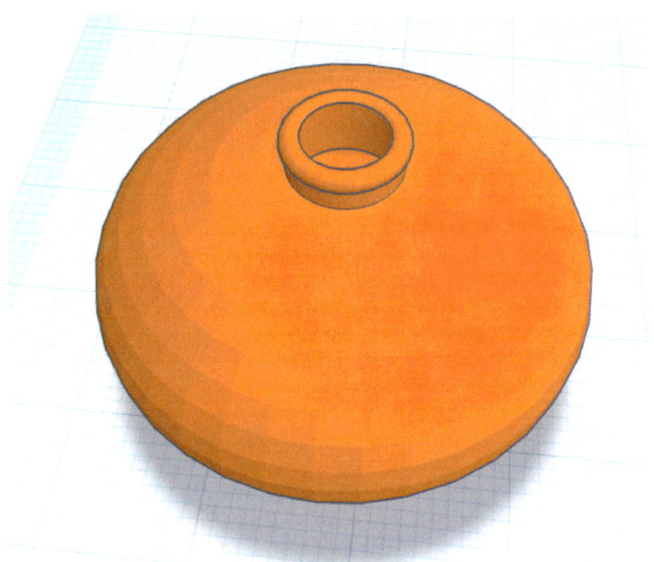

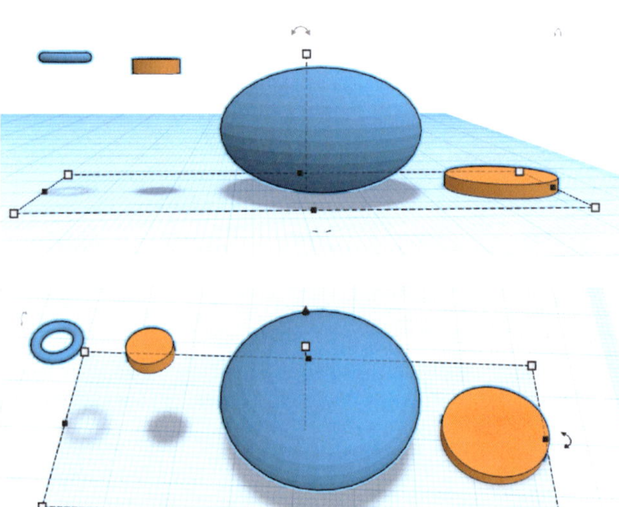

I have also done a version with a slight neck to it and a curved lip, which I achieved with a cylinder and a torus primitive.

The base is a cylinder 18mm diameter and 2mm high. The body is a sphere of 32mm diameter and 19mm high. The neck is a cylinder of 7mm diameter and 2mm high. Finally the lip is a torus of 7.65mm diameter, height 1.2mm and in the settings panel Radius of 1.5mm, Tube of 1.5mm.

Make smaller hole versions of the sphere and neck (at least 2mm reduction - trial and error will tell you what you can print on your machine), then align to centres on X,Y and group.

Boxes and crates

Fruit crates and boxes have slatted sides which would be impossible to print with filament because of the overhangs, but you can still print them in one piece by using a thin film of filament holding it together and a 45 degree mitre.

First decide your base size, I went for 26mm x 36mm (in grey with yellow outline).

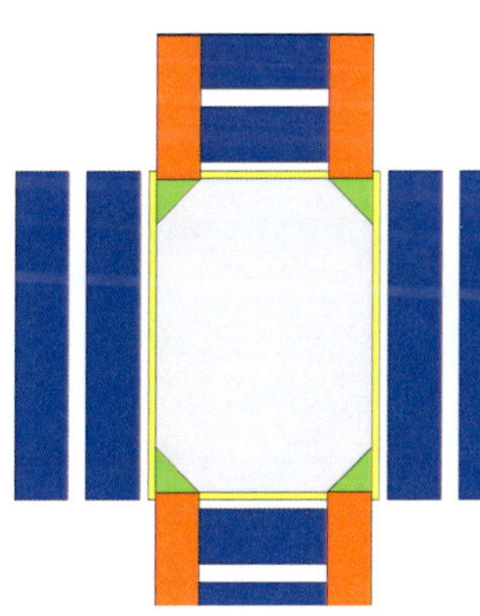

Then I put in corner posts of 5mm (in green) - export these corner triangles to SVG individually so we can rotate them later. I added matching 5mm strips (orange) to both sides of the support for ease of glueing.

I decided on side slats (blue) of 6mm. You can space and position these as you wish, adding more or less depending on the style you want, but be sure to space them evenly.

Import your corner post SVGs to Tinkercad and make them the height of your crate (mine is 16mm). Rotate them 90 degrees over to the horizontal to the left and right and raise them 1mm from the ground level.

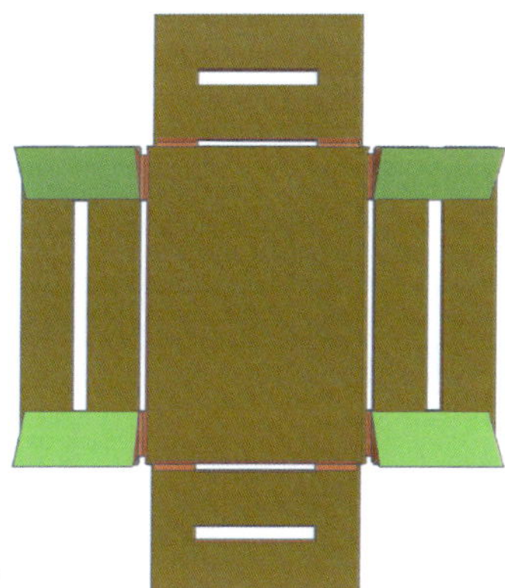

Import the base rectangle SVG and make it 1mm high.

Import the orange strips and make them 0.5mm high and blue strips making them 1mm high.

Align the orange strips to the vertical top of the blue strips so that when we fold them it leaves the effect of separation.

Group the base and strips together.

Now we make 2mm cubes of 38mm height (in red), rotate them down flat along the side of the crate, rotate them 45 degrees around the

vertical, and lift them 0.5mm from the ground level. Position them carefully so they create a perfect fold - this means the inner edge must be where you want the fold to finish upright. Convert them to holes and group to the base. This will leave a thin film of 0.5mm holding the crate together but in PLA it will be flexible

enough to fold up and strong enough to hold together. Repeat in the other direction. If you don't have a glueing jig you can use an elastic band to hold them in shape.

More shapes

You can change the base size and add different numbers of slats to make a themed variety of crates for a market stall or flower shop.

Next steps

You can use the flat pack approach on any difficult-to-print piece once you have perfected it here.

Brim, raft, supports and skirt

If you want to print a pole or similar curved shape in filament, there is very little chance it will stay on the bed as the print head runs around. The solution is to put **Brim** on, I tend to use 2-3mm for smaller objects. This prints one layer at the specified width all along the edges. It is easy to get off with scissors and a bit of sandpaper after printing, though if you have it in corners you may need a blade.

A **raft** is a disposable platform underneath the object, useful if the object has a small footprint on the bed for instance, or if it is delicate and difficult to remove from the bed of a resin printer. Often used with supports in resin printers.

A **support** is a disposable structure used to build over holes in filament that you can easily break off afterwards, or spindles to hold an item up cleanly from a raft in resin.

A **skirt** is an outline around and away from the print, used to prime the extruder. It may not be necessary if your printer, like mine, already automatically extrudes a little at the edge of the mat prior to print. However if you are in spiral vase mode it might be useful as there is very little to the base of the object and it might be weak if the extruder isn't quite fully engaged yet.

Shop table

For this one, to get all the measurements right in advance, I drew up a plan in a 2D vector program. You can get Inkscape free for this on most computers.

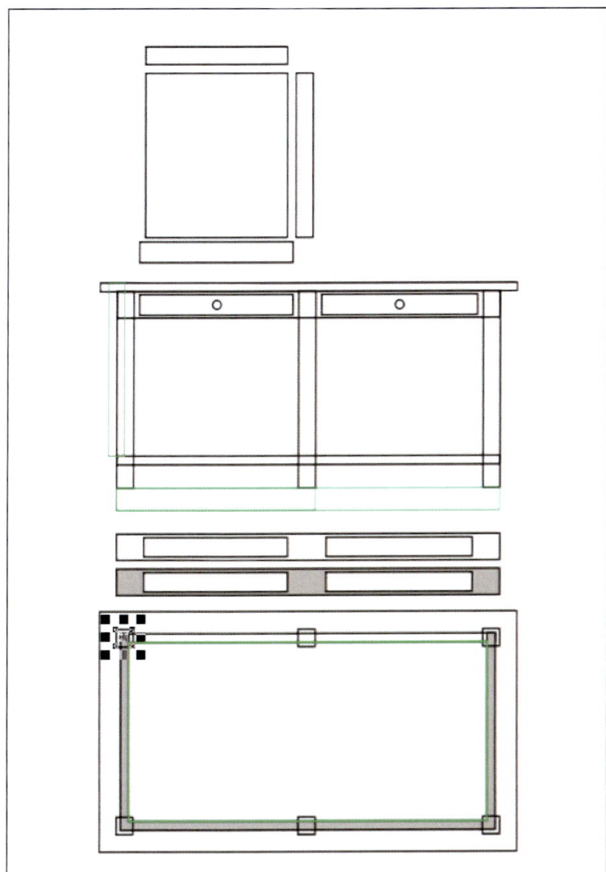

Back in your 3D software - for the main top make a cube of 82mm x 146mm and 3mm high.

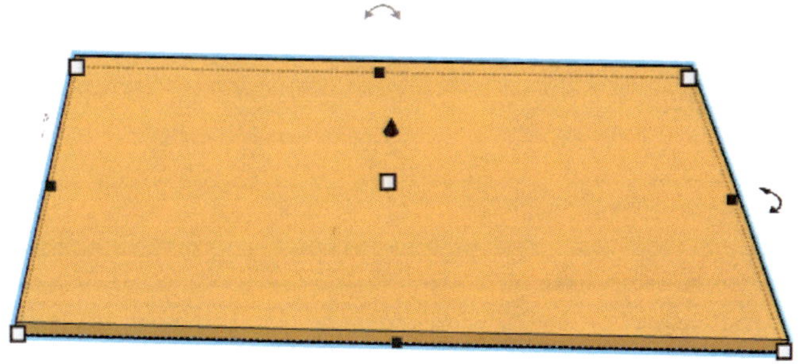

To make the legs make a cube of 6mm square and 67mm high.

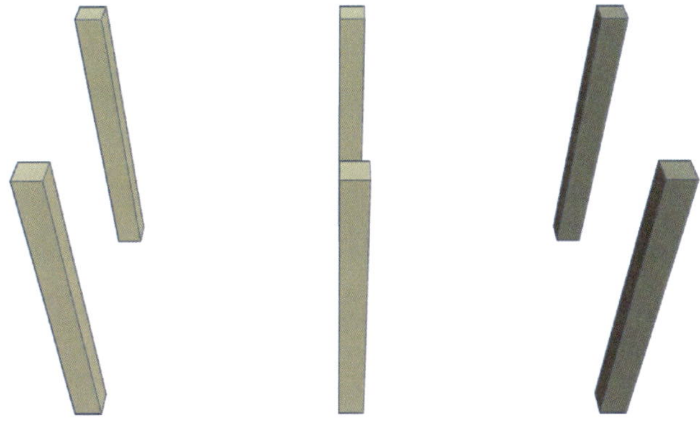

You will need 6 legs at the corners of a square 134mm x 70mm with 2 in the centre of the long side.

The infill pieces at the back and side are 3mm thick and 9mm high. You can draw these in by hand and adjust the width and height. If you don't want to add drawers, just put this infill at the front too.

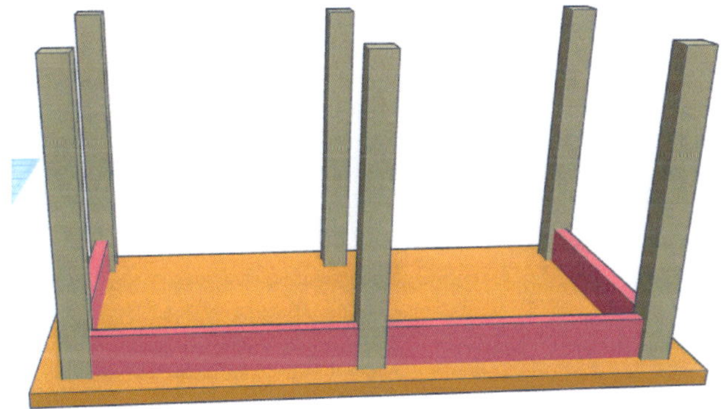

We will make the under-table shelf as a separate piece, a cube of 125mm x 61mm and 3mm high, and rely on a little flex in the filament to slot it into place at the end then glue it there.

To fit it to the frame, we will make a hole where we want it to sit, a cube 125.5mm x 61.5mm and 3.2mm high at a height of 59mm.

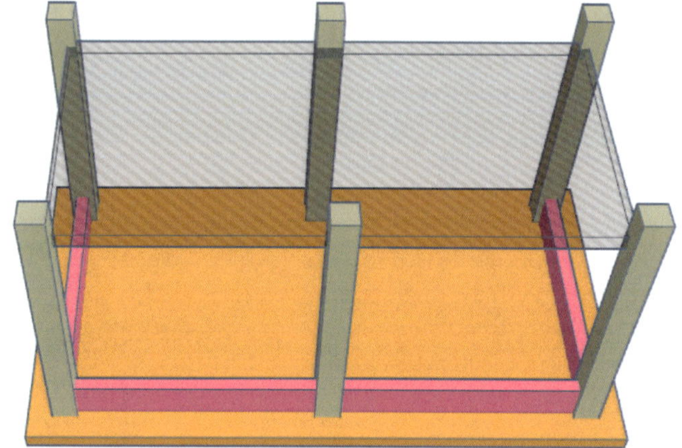

Make the front of the drawer as a glue-on piece so we can make it overhang and add a bobble for a handle. Make a cube of 54mm x 7mm and 3mm high.

Add a sphere of 3mm and raise it to height 2.5mm, centre it to the front piece.

Group up the pieces and export them to STL for slicing and print.

The frame for the drawers is a cube of 134mm x 9mm x 3mm thick. We will put 2 holes in it of 50.5mm x 6.5mm. Set them 9.75mm from each edge and centre them. You may want to put a copy of this frame 2/3rds of the way towards the back to support the drawers.

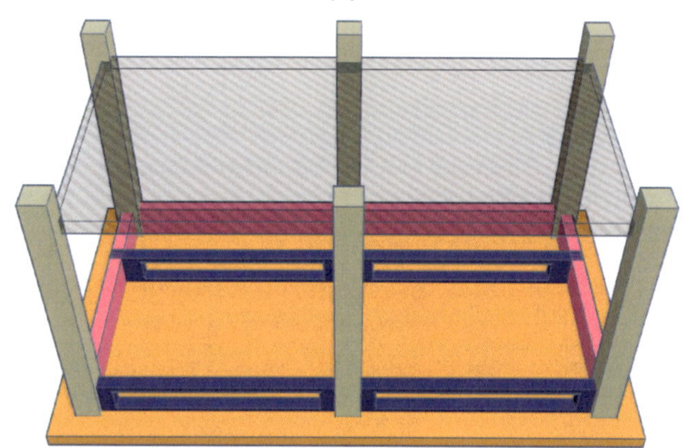

Now for the drawers, make a cube of 50mm x 56mm and 1mm high. Add cubes of 56mm x 1.5mm and 6mm high to each long side, and add a cube of 50mm x 1.5mm and 6mm high to the back and front.

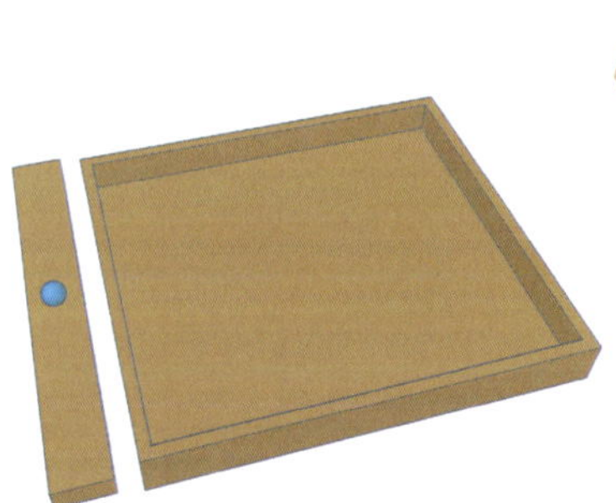

Hanging basket

Note that on the screenshots for this the basket pieces look significantly more chunky than the actual print it creates.

A nice simple one, just make an oval in Inkscape or similar, make a smaller copy inside it, and cut it off midway. Make sure it

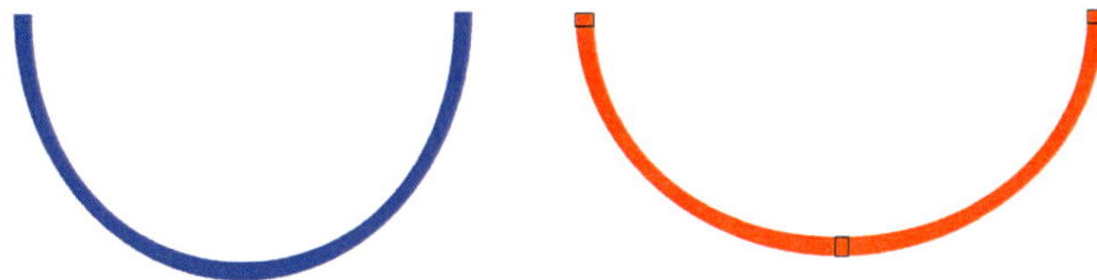

has at least 1mm thickness or you will have problems printing. On the diagram you should see I made 1mm boxes and pulled the shape to them. I went for approx 27mm diameter for the curve. Now Export this out to an SVG and bring it into Tinkercad at 1mm to 1.5mm height. Rotate it 90 degrees upright and pull it up to ground level.

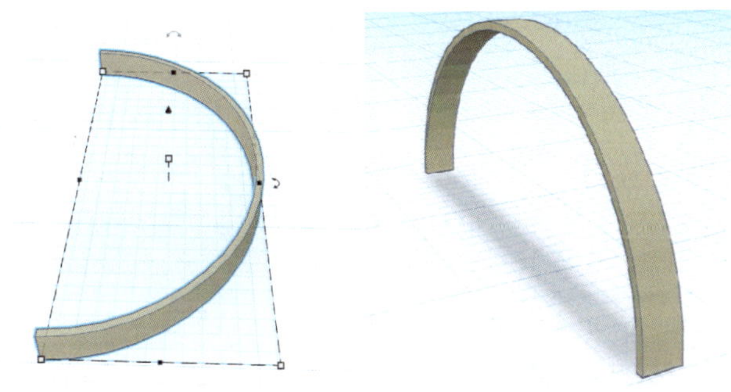

Click in Tinkercad to see the bottom measurement and make a cylinder 1mm bigger and one 2mm smaller as a hole, the height is your decision but I suggest 2mm.

Now copy and rotate the initial SVG by 45 degrees until you have filled the circle. Group all the SVG hoops, align to the centre of the

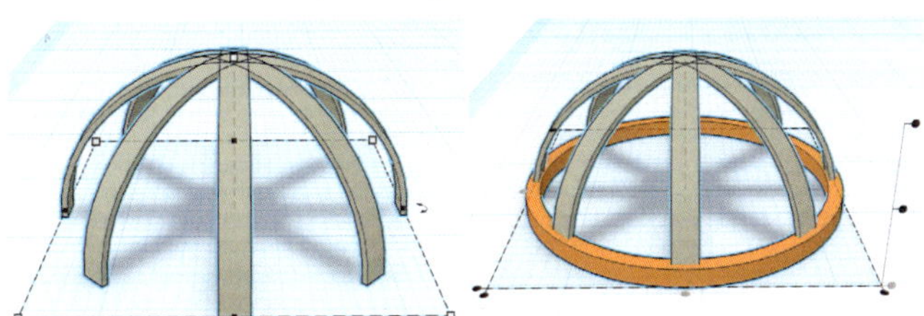

cylinder and group. You now have an easy hanging basket.

This is one to print in resin, or slowly and in highest detail setting on filament printers.

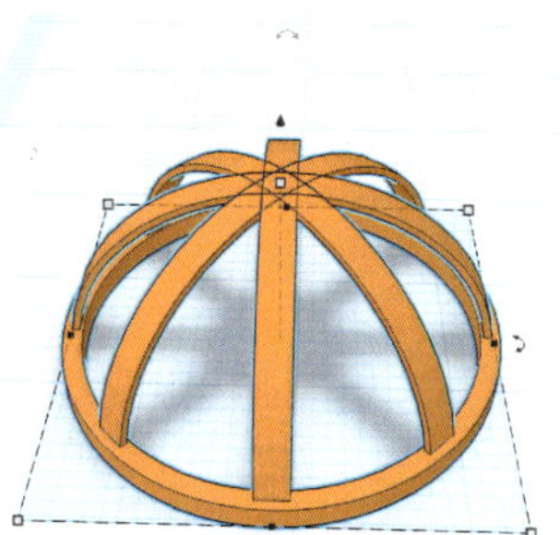

Another shape

I did a slightly different shape a bit deeper and had 4 hoops at 45 degrees, then 4 rings of narrowing diameters, the bottom 2mm high and the smaller ones 1.5mm high. Position the rings by eye so that they are touching a hoop at the base - this gives them something to "stand on" as they print and bridge.

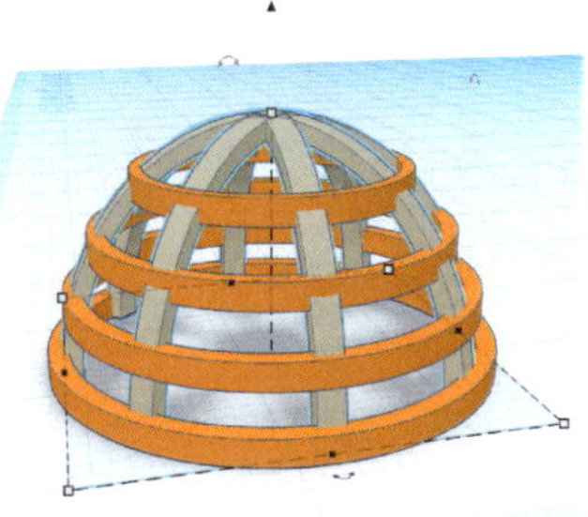

Next steps

You can add decorative elements inside the hoops like leaf shapes etc. as long as you follow bridging rules - always have somewhere to connect to before building up a shape so either have a vine across below the leaf or build up a curve, or at 45 degrees from the side. Here I chose to use supports to print the complex shape. It can take experimentation to get the supports easy to remove and leave a clean result, but for working in miniature it's a good technique to practise.

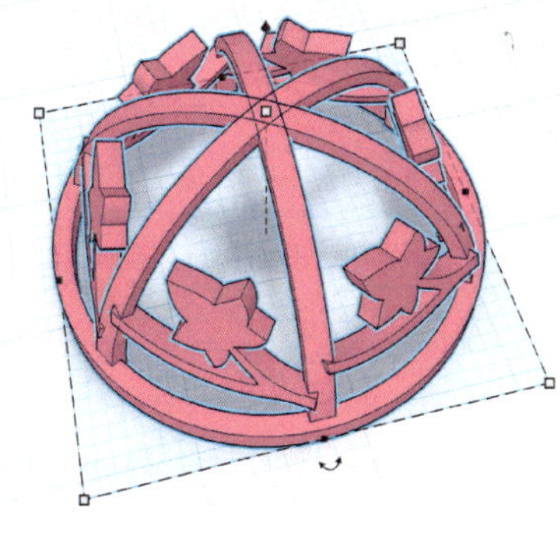

See the default automatically generated supports here (left) - it is quite a dense mess and the print is useless afterwards.

Some tweaking to make it "organic" gives a much better chance of getting a result. Maybe with careful scalpel work you could free the detail. Supports are meant to be detachable quite easily but working at this delicate level it is a challenge not to damage your work.

All that said I normally avoid using supports if I can engineer a way around it.

Ironwork (SVGs)

When designing the hanging bracket it is easiest to do some designs as SVGs.   First make a simple rectangle, copy a rectangle inside it by 1.5mm or so and cut it as a hole leaving a triangle. Next add a bar joining most of the way along the short side to the corner of the long side.

Finally take a circle and make a hole in it then cut it in half then, and join this crescent moon shape you just made to the corner of the long long side.

This leaves you to decide a pattern for the inner support, Angie drew this one and I made it into an SVG.

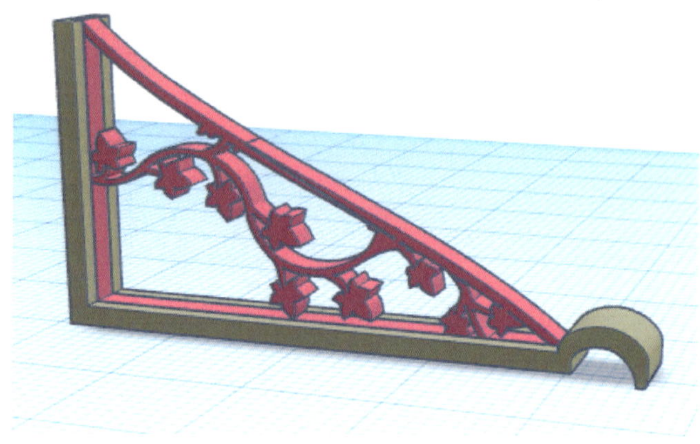

Note: be very careful lifting delicate designs from the print bed, it may be best to start at the strongest corner.

Improvement

You can use any pattern you like then add it to the angle bracket, above is one with a vine leaf design. If you want added subtlety you can print the main part of the bracket without a diagonal support first, with a groove to add in your pattern which is how I designed the Ivy one.

Hooks: for all of the above you will need a hooking system to attach the basket to a hanger. Jewellery makers have a lot of handy clips, hooks and miniature chains.

Part 2 - the shop front

Shop front

Now we have a rough idea how to make things, let's go the whole way and make a simple shop front. Once you have these principles you can add as much detail and design work as you like to your own shop. This will take a little time and patience and introduce a couple of new ways of working, but if you have followed the designs so far there is nothing actually difficult. I maybe wouldn't skip straight to this section if you haven't had some practice first.

To start the project I looked at lots of photos of shop windows and made up a diagram of a simplified one from a certain era.

This has Georgian style windows panes using "the golden ratio" of approx 1.6 tall to 1 wide. However the shop is from a later period as the window frames are out of this ratio, so they are more like a Victorian style. It's best at this point to create a simple SVG using our measurements or your own design.

The first thing to decide is the size, mostly English shops are built on 12, 15 or 18ft frontages, this one is 15ft (with an optional 3ft side door for a flat above taking it back to 18ft). This gives us 381mm for the width in 12th scale. Now for the height, I have gone for 275mm which is over 10ft in 12th scale but allows room for flooring and a ceiling if wanted.

I wanted a header board over the top for the shop name, 381mm x 46.5mm which I have made at a slight tilt downwards.

The windows are boxed out into the street and have side panels and supporting brackets with a bit of trim top & bottom. The main window frame is 125mm wide x 165mm high with 4x4 window panes. The side panels are 35mm wide with 1x4 window panes.

The door is 73mm x 161mm with hinges extra to that and added later. It has 4 windows above the letterbox and 2 below with space for a kick strip.

There is a 50mm window frame above the door. I have added a step to the door and a couple of slightly decorative panels below the windows which are in keeping with the style.

I chose to recess the door on a frame 7mm wide and 10mm deep holding the door and top window, with hinge holes built into the frame. This allows the door to move.

Finally simple squared columns are left and right of the shop front, made with 4 small diameter cylinders to give a textured effect. These start at the floor and stop at the shop sign which is just proud of them.

Window Frame

For some of these operations you will find it far easier to use a 2D vector program, e.g. Inkscape, as it has easy "Align and Distribute" and "Snap To" tools. Let's do an example.

Make a rectangle 125mm wide x 165mm high. Inside it put a rectangle of 25.5 x 33.5 and move it 5mm right inside the larger rectangle,

and 7.5mm down. I have used little green blocks to mark this out, and it also allows me to copy it to the next row or column using "snap to" (which lets you magnetically click to objects, centres, edges or guidelines as you choose).

Copy and paste it 3 times. Move one to the right edge (minus 5mm). Select all 4 then use Distribute -> Horizontal Spacing (the wording may change depending on which software and version you have but the functionality is the same). This will give you a perfectly spaced row of window panes. If any are out of line vertically you can select it, then select the correct one and Align -> Top. The order of selection can be important.

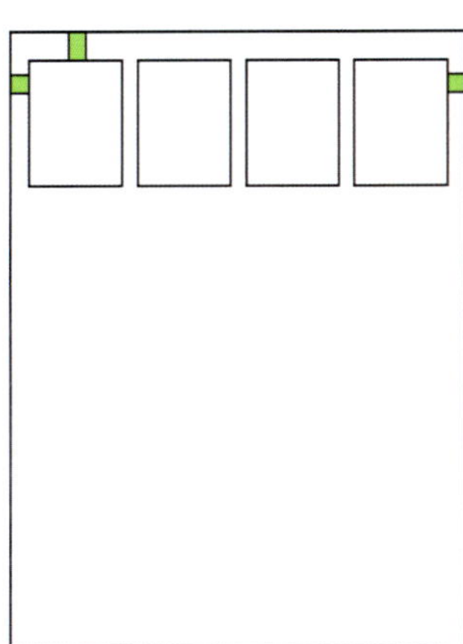

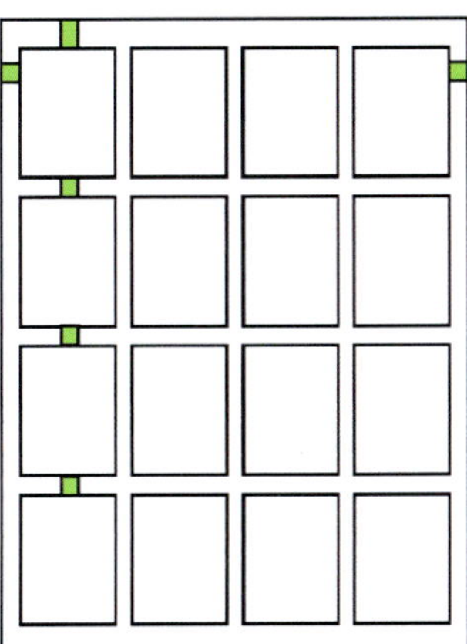

Normally the last selected item is seen as "correct".

You can now group the 4 panes, copy and move to 5mm below the first set. Repeat this until you have 4x4. Group all 4 groups of 4 and select the outer frame, then Align -> Centre Horizontally and Align -> Centre Vertically. It is now perfectly spaced inside the frame. You can remove the guide pieces (green in the diagrams).

You can now do the final pieces to make it into a shape: cut the front (window panes) from the back (large rectangle). Then export the result to an SVG.

To make the side panels the simplest way is to draw a box over the first column of panes and use

Intersect. The language of vector graphics can be a little complex unless you already know. I had to explain this command very carefully to Angie. The intersect tool will make a new shape from previously constructed parts and will only contain the area that coincides between the two. Think of a Venn diagram with two intersecting circles. The middle part is the intersection. If you look below at the three circles and the square around them. The green part is where they intersect. Now see the red box we have drawn round part of the window. This part is the 'intersect'-ion and we can then copy it to make the side of the bay window.

Don't print these frames yet there are a couple more stages first.

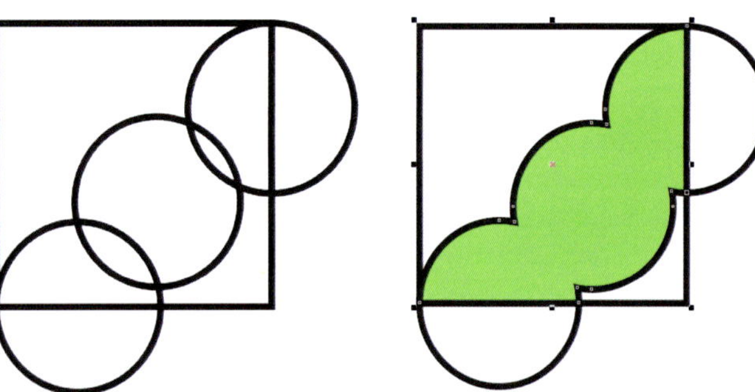

Top and bottom blocks

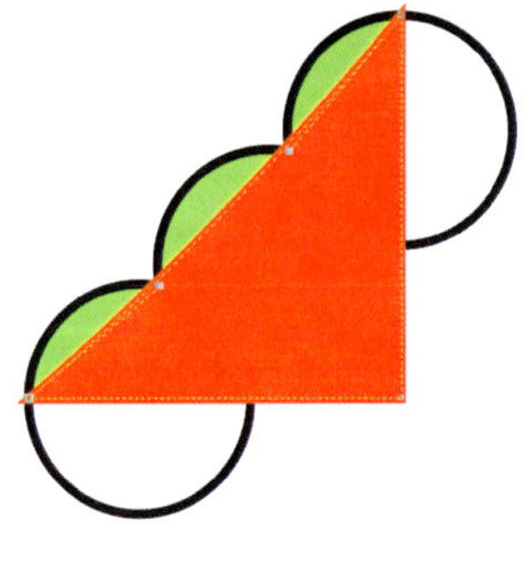

I have made blocks to fill in the top and bottom of the window boxouts, with a decorative architrave around them.

To make the architrave I went into a 2D program and drew a box with the dimensions I wanted - 3mm square - then put 3 circles in it and arranged them until it gave me a profile (green). I will use it to make an extrusion.

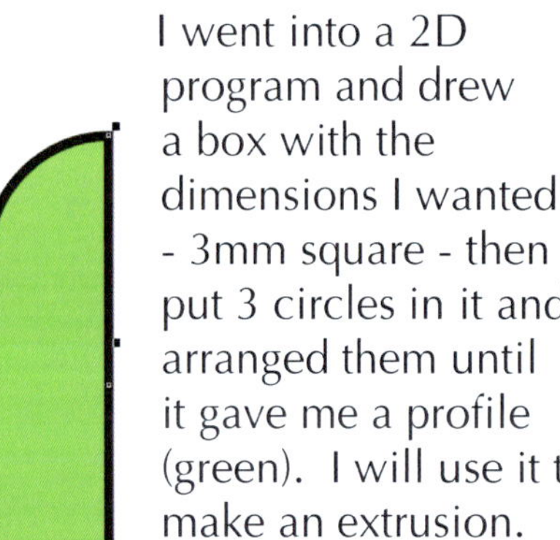

I Weld-ed the circles and then Intersect-ed them with the square.

Finally I drew a triangle from corner to corner through the circles to make it into a simple architrave.

Export this finished trim as an SVG into the 3D software and you can extend it and mitre it as you need.

In the 3D software I made a block (red) of 125.5mm x 35mm and 3mm high. I ran

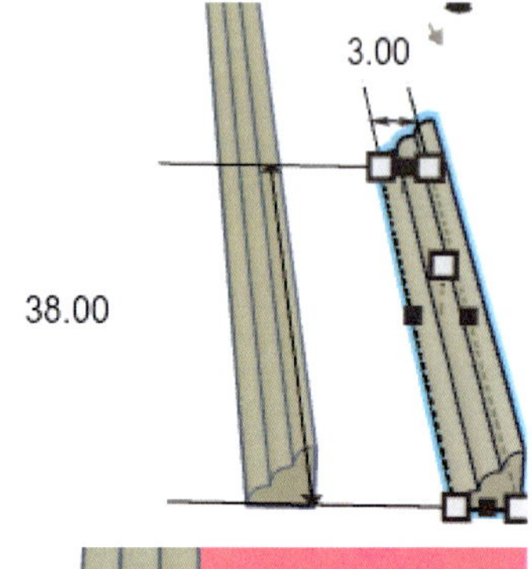

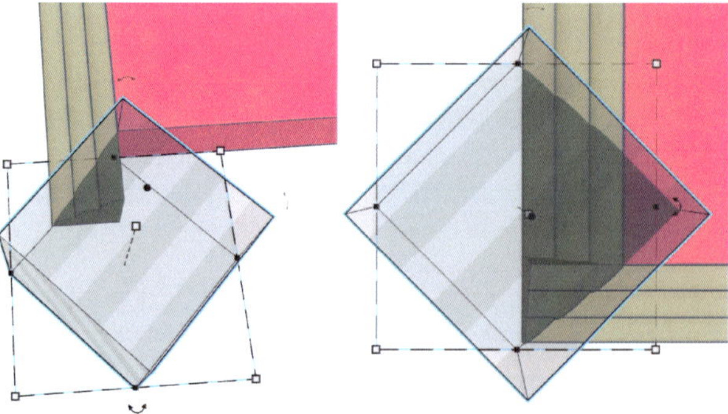

a piece of the architrave to length 38mm at the side, rotated a copy and ran it 131.5mm along the front. This leaves a weird shape at the intersection where we need a mitred corner. To deal with this we can make a hole cube, rotating it to 45 degrees and grouping the hole with the side piece, effectively creating a mitred corner. Repeat with the base

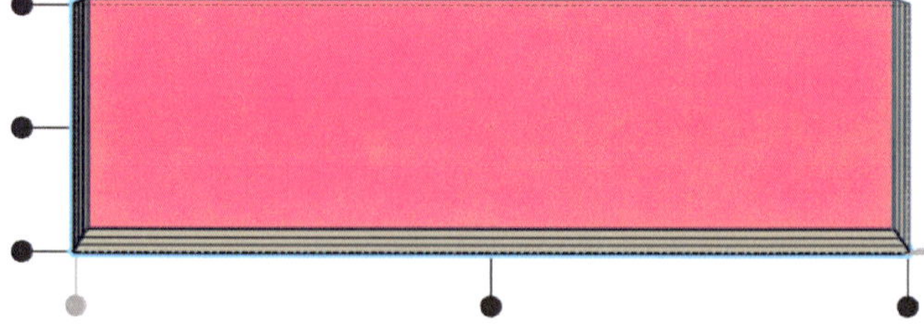

piece. This takes perfect alignment so zoom in closely. Then 'mirror' a copy for the other side of the window.

Supports

Here I am making brackets to go under the windows as I have seen this on real shops. Make a rectangle 35mm x 5mm, then add a rectangle 5mm x 33mm to the left and under it.

Now to make an easy curve I have drawn an

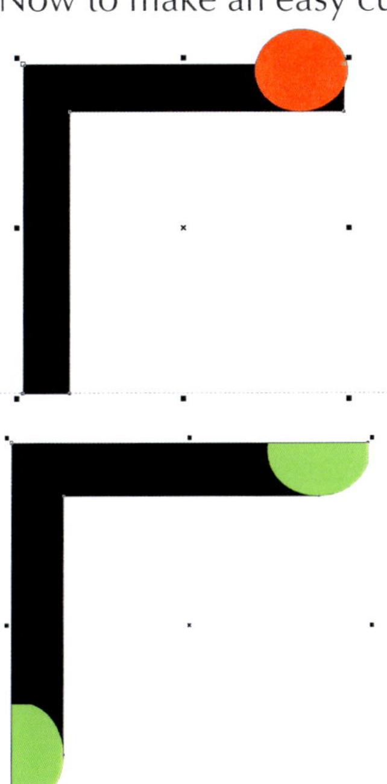

ellipse (in red) so that it covers the top right corner and contacts at the bottom. Using Intersect I can make the piece that joins it perfectly (in green) and remove the original circle and the corner of the rectangle. Copy the intersect down and rotate it to the bottom corner.

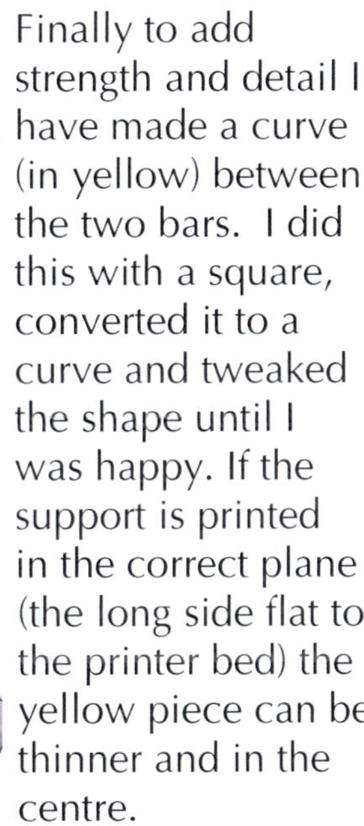

Finally to add strength and detail I have made a curve (in yellow) between the two bars. I did this with a square, converted it to a curve and tweaked the shape until I was happy. If the support is printed in the correct plane (the long side flat to the printer bed) the yellow piece can be thinner and in the centre.

Door

Make a rectangle of 73mm x 161mm. Inside it place a rectangle of 28mm x 37mm. Space it 5mm in from the left and 6mm in from the top. Make a copy 5mm away to the right. Copy these 2 panes 5mm down.

For a letterbox make a rectangle 24mm x 9mm and place it 5mm below the pane. 5mm below this add 2 more panes. This leaves a space underneath for a kick strip or runoff.

Group all the panes and Align -> Centre Horizontally to the large rectangle. Then do the same for the letterbox.

We will add the hinge pieces later in the 3D software as it requires perfect alignment with the frame, which we have to assemble there to get the required holes for the hinge.

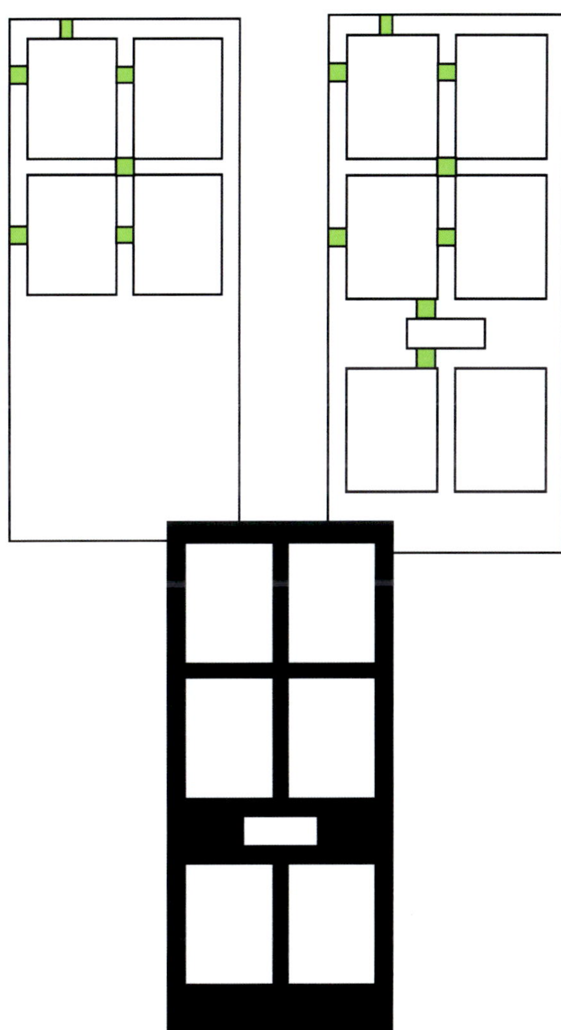

Splitting for 12th scale

As the shop front is 381mm x 275mm that is too large for most filament printers to print in one piece, except in 24th scale which works quite nicely. I am going to break it into sections.

First the shop sign across the top can be cut in half to 190.5mm x 46.5mm, and printed twice. If you want you can build the name into this piece when printing and then paint it or print it in a different colour.

Next the side section of the shop can be made from a rectangle of 153.5mm x 228.25mm. If this is still too big for your printer it can be split further at the lower part of the window opening. We will only make this once then mirror it horizontally.

Finally the door frame will be 87.5mm x 228.5mm. Note the uprights on this will overlap the side panels internally (where the green line is on the side section diagram opposite). This makes for a strong bond.

Shop sign / fascia

This looks best with a slight tilt downwards. The easiest way to achieve this is to make an SVG of the shape profile we want and export that.

Make a rectangle of 46.5mm x 13mm, then another of 46.5mm x 5mm. Convert the larger rectangle to a curve (or path, or whatever your software calls it) and use the edit tool to move one corner downwards to meet the smaller box.

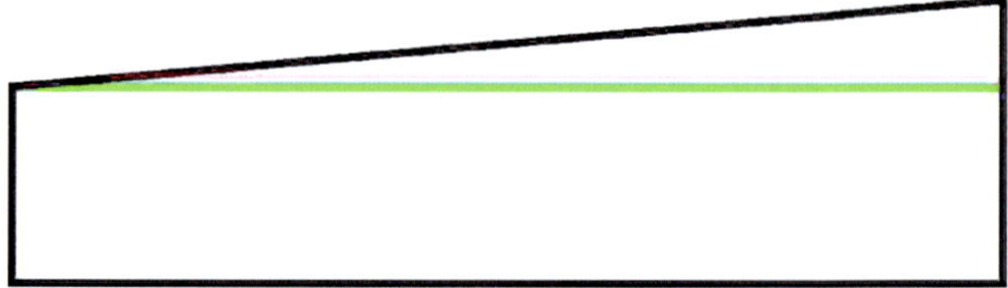

Export the shape as an SVG, import into the 3D software and rotate it 90 degrees so it is "on end".

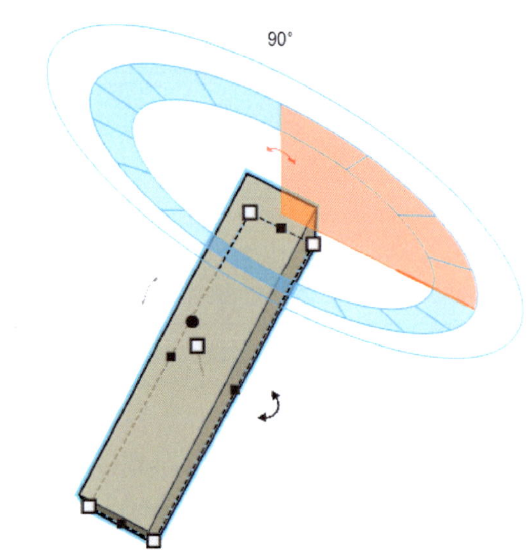

Now adjust its length to 190.5mm.

This fascia is going to have architrave added later as we design the corbel … OK, translation: we are going to add fiddly bits top and bottom as we do the top end of the column.

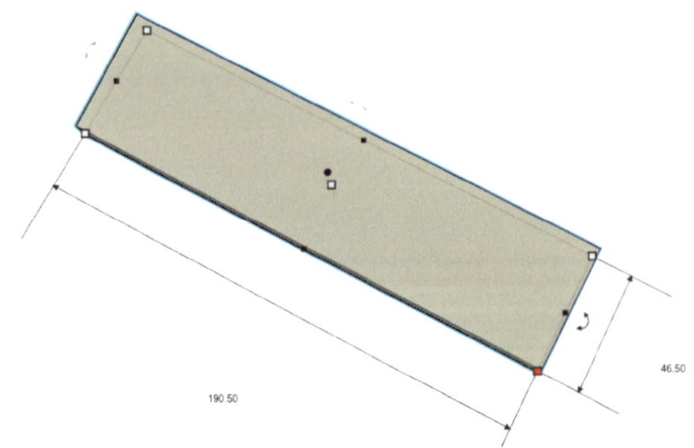

Side section

We are going to put a squared column at one side of this and decorative panelling underneath the window.

At this stage it is best to make sure we leave room around the panelling for the brackets holding the window and the bottom of the window.

We can centre the opening and panelling in the space left after the column is added. This is slightly different to the very first diagram, but best to get it perfect now before we print - see the blue lines.

To leave space for the column draw a rectangle of 21.25mm wide the full height of the side section and to the left.

The opening for the window will be 115.25mm x 150mm. This is smaller than the window box and leaves a contact area to glue the frames to the shop front.

For panelling I have made a simple decoration of 3 nested boxes of 110mm x 33mm at 6mm high, 104mm x 27mm at 5mm high and 97mm x 20mm at 4mm high. Feel free to modify this to your taste.

Columns

I am decorating the column using 4 cylinders, and with a solid plinth at the base up to the bottom of the window level. Always align things to something else when designing from scratch, it makes it look more cohesive.

To make the plinth you can go straight into the 3D software, make a cube of 13mm high, then shape it to 21.3mm x 48.5mm.

Next make a cylinder of 8mm diameter and rotate it on its side. Copy 3 more and align them just touching each other.

Corbel

Now to think about the top and make a plain, curved, corbel. And to be awkward I am going to think of it in 2 planes at once. First of all the "footprint" of the corbel is the column width 21.3mm and the fascia (shop sign) height 46.5, so that is the grey box. Next I am going to add a simple 3 steps (green boxes) to it that overhang at the side. Each is a box of 1.5mm high and 1.5mm extra width added at each stage. When we bring it into the 3D

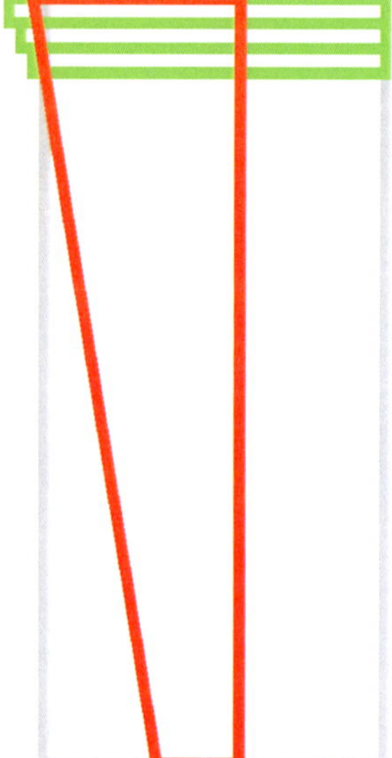

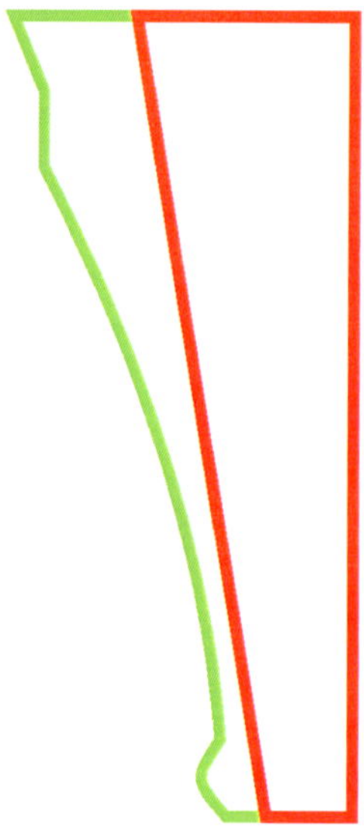

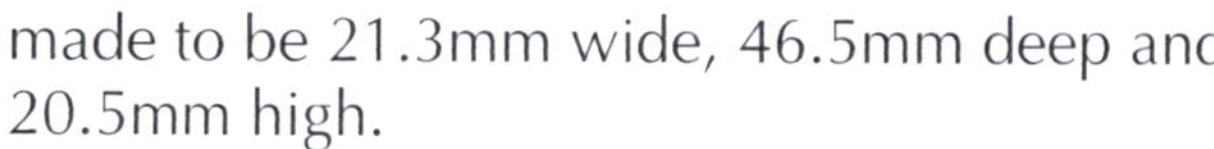

made to be 21.3mm wide, 46.5mm deep and 20.5mm high.

Next bring in the three steps, make the smallest 19.8mm high, the next 21.3mm and the largest 22.8mm high. Align them to the right of the main piece.

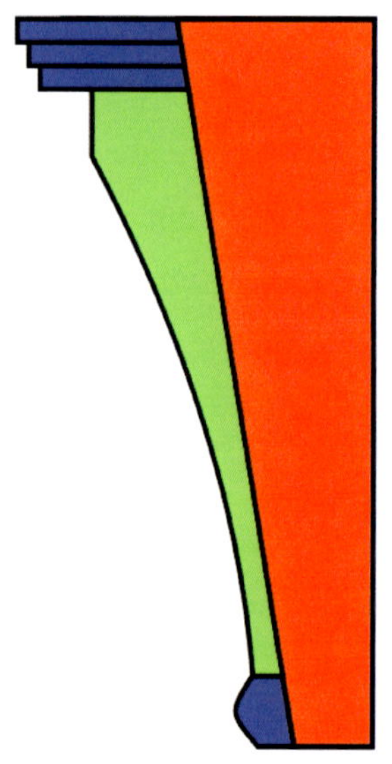

software we will also increase the height in the same way, 1.5mm steps, so it has the same effect as at the side.

Next I bring in the profile we made for the shop sign earlier, and adding a profile to it that looks right to me. It's all your choice at this point.
I have a curve at the bottom which I have Intersect-ed so I can add it all across the sign and the column.

Import the corbel shape and rotate it. It should be

For the sign, bring in the profile and the bottom curve. We will stretch the 3 steps onto them later.

Rotate it and size it to 190.5mm long. Put this on top of the column and align it to the left. Put the corbel on top of the column but align it to the right, so the tiered steps overhang at the edge.

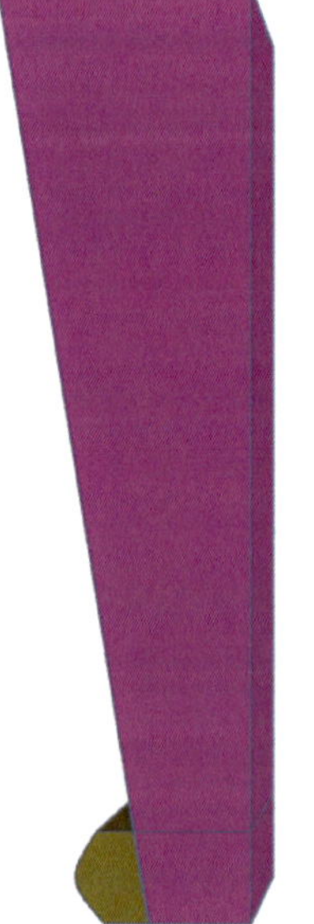

Door and frame

The door frame is a solid support for the door to swing on, to glue the sign and sides to and also have a step coming out into the street. Finally the top will have a small strip so it appears flush to the side pieces in between the two verticals.

We will start by importing the outline of the door frame from the master diagram as an SVG (grey in the screenshots) and making it 5mm high. Also bring in the top window piece at 2.5mm high. We will later add "glass" and a 2mm high backing piece (blue in the screenshots) to this to bring it out flush to the cross pieces.

Choose which side you want the door to open at this point by choosing which side to hinge. Be aware that we are printing the door frame upside down so it will be reversed on screen. My example hinges at the right (on screen) so it opens at the left.

This is also a difficult task as the post and holes have to align perfectly with the hinges on the door, and be shaped to allow the door to open smoothly. Make this bar of the frame out of a rectangle (yellow

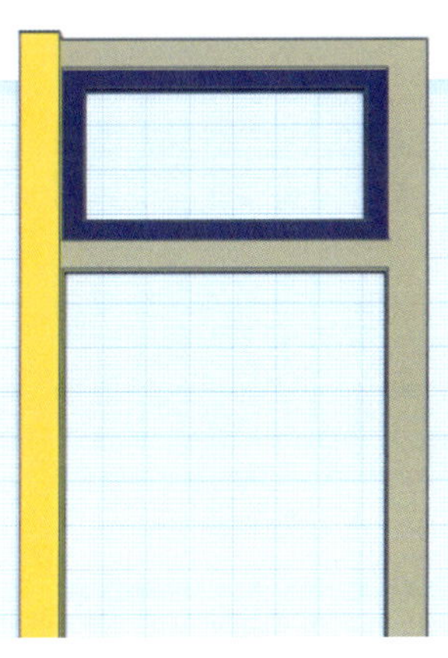

in the screenshots) initially 10mm high, and align a horizontal cylinder (orange in the screenshots) of 8.5mm to the top of it.

This needs to run down into the frame where you want the door to sit. I also do a rectangular cutaway into the frame to allow easy insertion. Line the door up perfectly into the opening of the frame and we will assemble it all in place here.

For the cutaway make a cube 12.5mm x 8.5mm wide and 8mm high. This should allow the door to open and close with approx 90 degrees of movement.

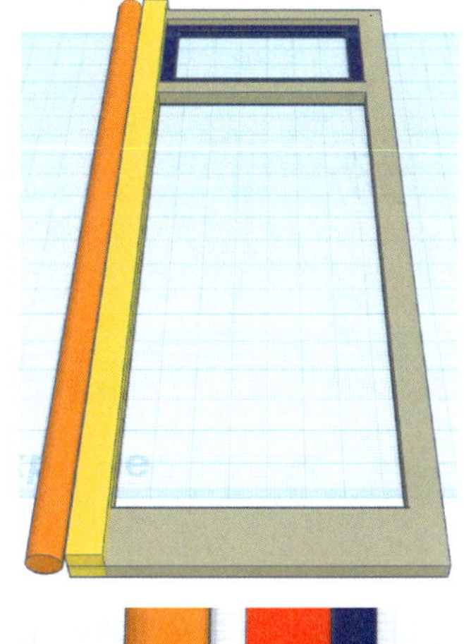

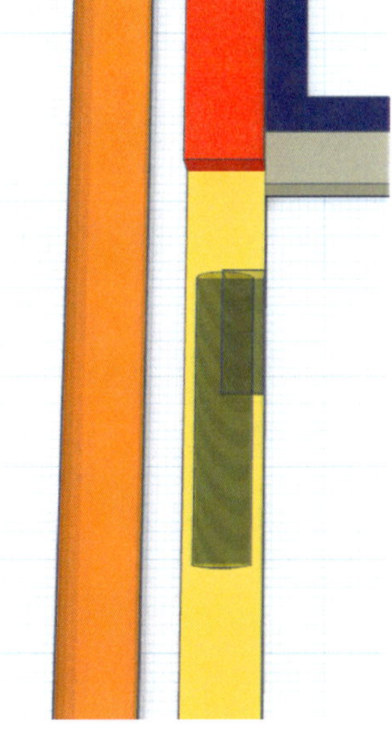

Make a cylinder 6mm diameter (slightly larger than the 5mm pin we will put on the hinge) and 18mm long. Align the cylinder to the upper edge (relative to the door frame) of the cube and so the centre of it touches the top of the cube vertically. Make the cylinder and cube into holes.

Then reduce the initial frame rectangle to 5.75mm and it should leave a perfect half circle above it.

Pad out the kickboard and top window sections of the door frame with cubes of 8.5mm wide and 10mm high to strengthen the frame and to make it look right. Pad out the

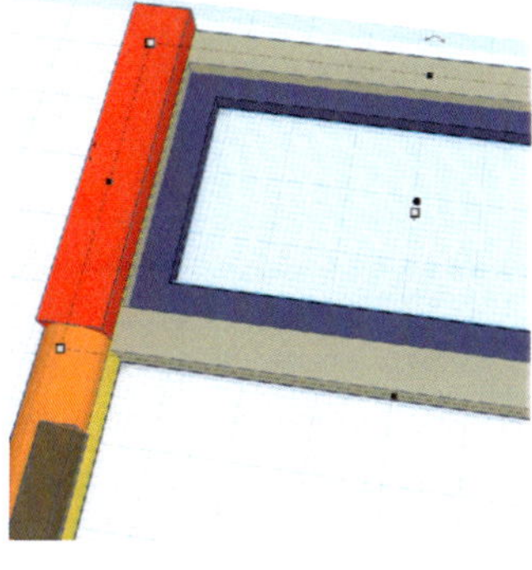

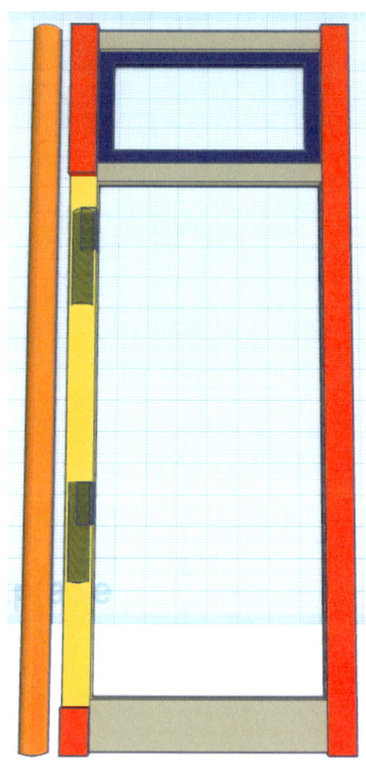

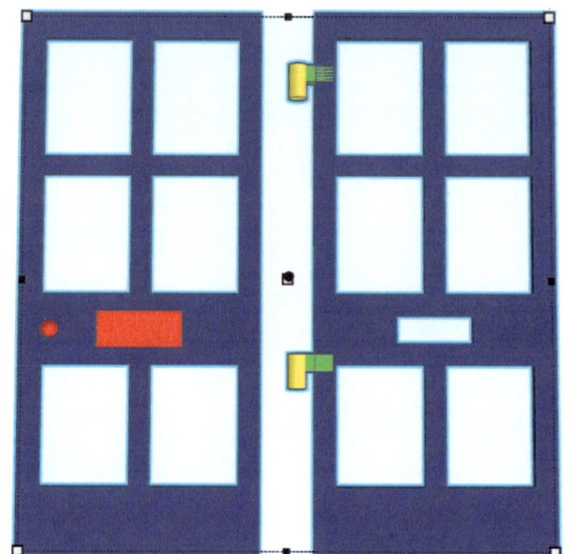

entire other side of the frame to 10mm high too. These pieces are in red on the screenshot.

The hinge on the door will be a cube (green) of 5mm x 11mm and 3mm high. Align this to the bottom edge of the cutaway cube (relative to the door frame).

Add a cylinder of 5mm diameter (yellow) and length 10mm, align this to the top edge of the hinge so it hangs down and into the circular hole we made in the frame.

Repeat these pieces in the frame and on the door at the second point you want to hinge the door. You can add a letterbox cover and door handle (both red) before or after print depending if you want to use metallic for these.

Group all the pieces in the door frame, then on the door.

Architrave for the door and frame

Use the architrave we made earlier at its original 3mm x 3mm to make surrounds for the door frame and inner windows. This time it is easy, the pieces will align straight together with no need to mitre. The top window for the door needs the backing piece cut down to 73.3mm by 50.15mm and 2mm high. Surround this on all 4 sides with the architrave at 3mm high sat on top.

The architrave for the door frame should be 3 sided measuring 73.3mm by 162.2 mm and 3mm high.

Doorstep

A simple box extending out from the inner door frame into the street.

The step is a Cube primitive of 73.8mm by 25mm and 15.48mm high.

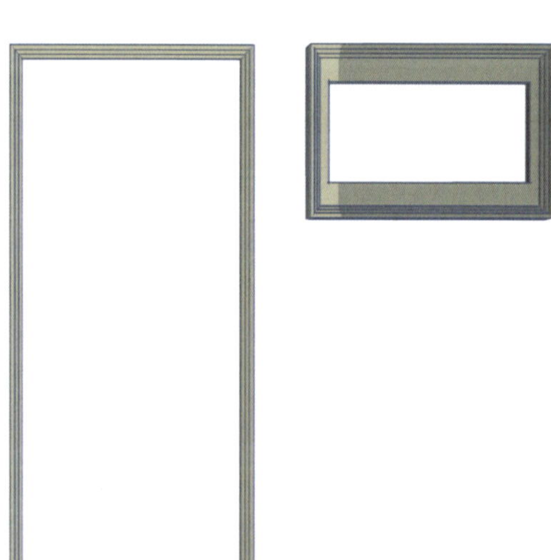

Name

For the shop name we went with Mrs Dry, Draper from a shop Angie remembers as a child. Draper is the same as haberdasher.

I made a 1mm high surround the size of the space available minus a small amount for a snug fit: 337.9mm by 36mm. CAREFULLY measure your own gaps after assembly! I chose a serif font in all caps for the writing and raised it to 3mm. Printing on this will be tricky depending on your printer's support for colours or ability to switch filament mid print. My intention was a black surround and changing the filament by pausing the Geeetech as soon as it passed 1mm height.

After making this up in Tinkercad and grouping it I split it into 2 sections to fit my printer bed, which is quickly done with a hole cube allowing you precise control to split it between letters.

Fixing a mess

I wanted to print this in two colours: black for the baseboard and silver for the letters. On most printers this changeover is simple and can be chosen in the slicer, but at the time of print only the old Geeetech was available, so at the height it was ready to switch (1.3mm) I changed the filament manually.

The machine then proceeded to drop the head heat down to room temperature and crash entirely. There was no option to resume on reboot. So at this point I realised I would have to do a complex workaround to save the time already put into the print. I had to go to

Prusaslicer and export the final layout as an STL, go back to Tinkercad and cut the base layer off, re-export just the top 1.8mm of the letters and finally choose Printers -> General -> Z Offset 1.3mm and Printer Settings -> First layer height 0.1mm. I re-sliced the file and crossed my fingers.

I had forgotten that the Geeetech puts a Skirt around the whole print so I (well OK, Angie in the end) had to hack off a silver thread after printing, otherwise the fix worked. If there's a next time for this method I will try to remember to turn off the Skirt.

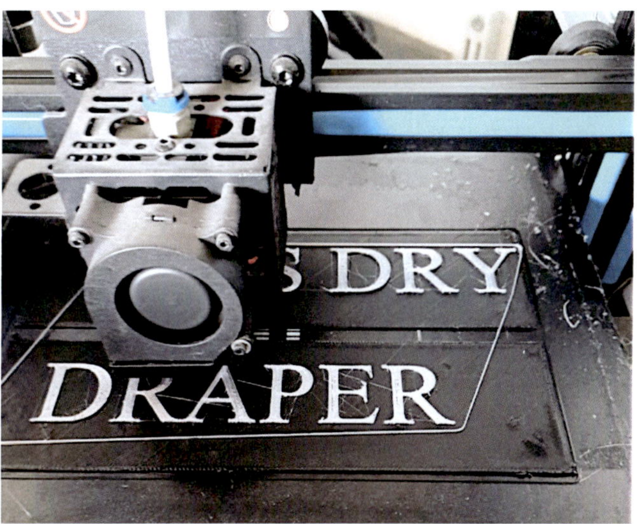

Of course that kind of mess can't end easily, I took them off the bed too soon and they are very thin and so it caused warping. I had to warm the bed again and put them on it for 5 minutes with things placed to flatten them down.

Finishing pieces - "glass"

Now we have all the main parts we need to think about adding glass to them. I have made the frames and door quite thin to leave room for a backing piece. In the backing piece will be a 0.5mm recess to put the acetate, plastic, glass or whatever you decide to use.

If you make the backing piece for the frame or door 2mm thick you can draw a box slightly bigger than the glass and make it 0.5mm high. Then align it to the vertical top of the frame or door. This will then print with a recess. Note the size of the recess and cut the glass to 2mm less for ease of insertion.

For the main window I have cut pieces of acetate 119mm by 58mm and for the side panels 32mm by 158mm.

For the bottom of the door 44mm by 68mm, for the top of the door 83.5mm by 68mm and for the window above the door 44mm by 68mm again.

For the glass display counter I have cut 142mm by 68mm and 142mm by 56mm.

Wallpaper etc.

The finish on the bed facing side of the shop front might be rough, you could sand and fill it or I have made SVGs for print or cut for the inner. With this you can give it a white finish, you can just cut this on card or use any colour or pattern of your choice. Or even print it in the filament you like to match the rest.

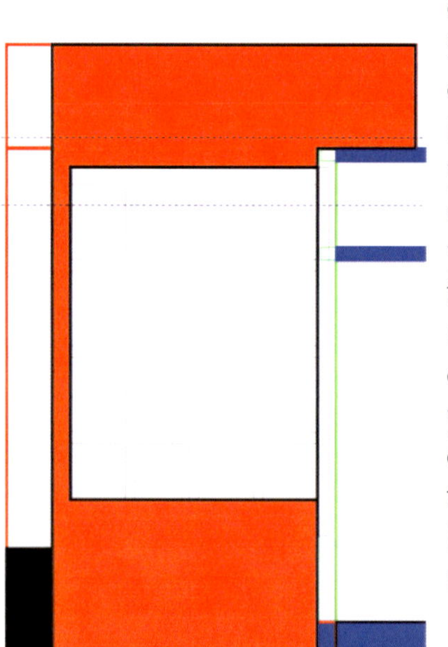

Here I have drawn around the red area that needs to be backed and taken it back 0.2mm at each corner.

Skirting board

Using the template for the door frame architrave (but less deep, say 3 or 4mm) you can run skirting boards across the bottom of the shop just to finish it off.

Simply measure the distances *after* finishing the assembly to give yourself some leeway, and attach all around the front and side - missing the column and doorways.

Assembling the window bay

First assemble in Tinkercad

Place the base frame with the hole for the window. Next add the kickboard and panel details.

Fitting right up flush to the hole at top and bottom add the top and bottom blocks - the flat pieces with architrave - with the thickest part towards the hole each time.

Add the brackets underneath keeping clear of the panel detail.

Now add the side panel window frames into the space - they should fit perfectly, maybe even overlap a little. If not go back and adjust the top and bottom blocks.

Finally add in the column and plinth to the side, it should all be symmetrical and it doesn't matter which side as we will mirror a copy when it is done.

If you have a large bed or are working in 1/16th or 1/24th you can also add the fascia board and column corbel to the top at this point. Otherwise keep this pair as a separate print.

Keep the main window frame as a separate slot in piece.

After putting all this to a test print I found the main window frame as I gave the measurements doesn't quite fit top-to-bottom, maybe 1mm or so out. I went back to the design, rotated it up and placed it against the hole it should fit in. Yes, not quite. But I can

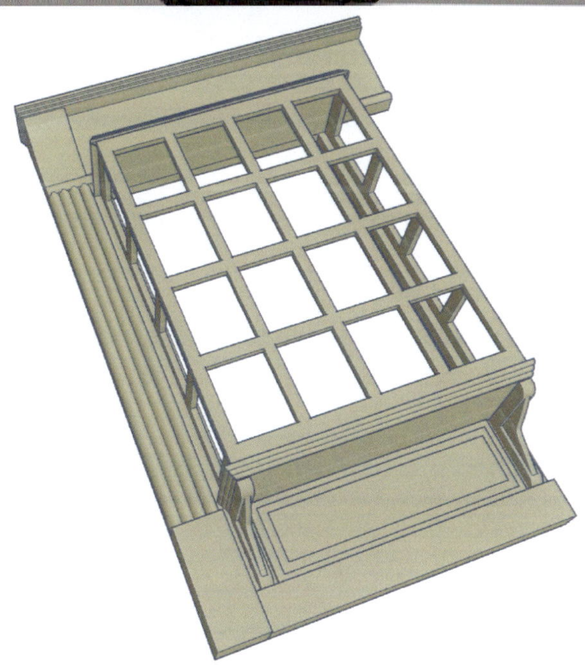

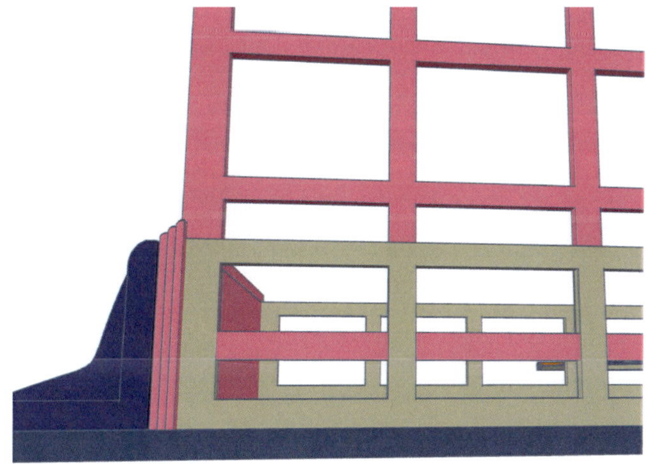

trim it here in place to be exact with a couple of carefully aligned hole blocks. Align them to the edge of the hole and move them 0.25mm in. Note this will also mean the slot-in side

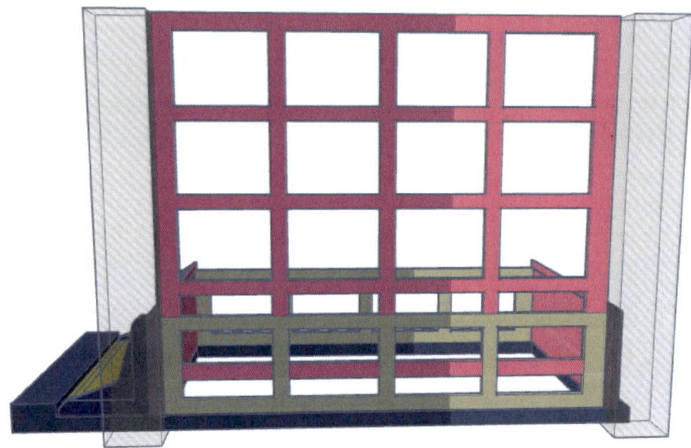

panels, there to hold the acetate, will need to be trimmed too, so I will re-create them using this corrected frame and cut another .25mm trim each side, to fit into the bay snugly.

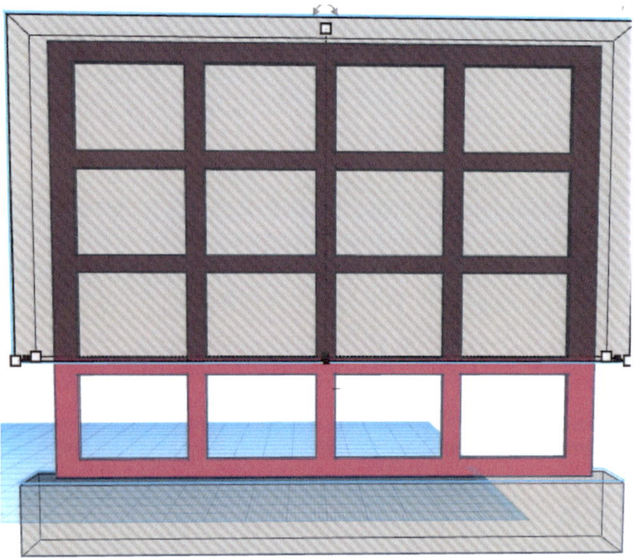

Yes I could go back, make everything perfect again, probably having to go right back to the original SVGs, but it does no harm to just fix what you have when you are this far in. The window bay looks good, so I went with it. All that matters is they perfectly align with the finished bay. This is how to fix small measurement errors in Tinkercad. There are always errors and last minute design changes so no need to panic if something isn't perfect the first time.

As always at this point I ran into a problem - a quick fix? Computers don't allow them. I grouped the frame and the holes and … error. Tinkercad has a workspace boundary that it only subtly warns you about, you can move

things past it for tidiness but group, export etc. don't work there. See the shadow at the left of the shop front above. I didn't want to move around my assemblies too much as I can drop SVGs perfectly in place over them from my main drawing to add new pieces. Eventually I figured out I needed to take the frame and hole blocks out to a new, clean file. Grouping worked as expected in the new file and I copied the trimmed pieces back to my main assembly.

At this point I did fast, 24th scale, test prints and it looked like the project was on track so I made full size versions.

Backing pieces

The main window frame and side panels both need slghtly cropped versions to stick inside the window box and hold the glass in. Ideally you will follow the same procedure to align them in Tinkercad and trim them to fit, plus an extra bit of trimming for luck as always (e.g. 0.25mm). I arranged mine so the main window backer is neatly inside the box sitting on the two side pieces so the side pieces need even more trimming on the outside edge.

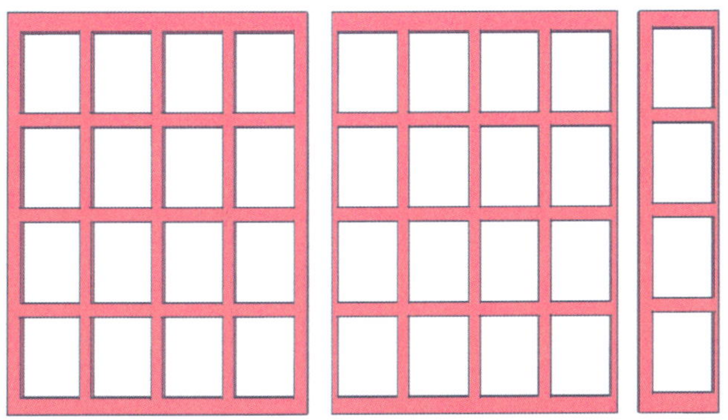

Finishing pieces - "glass"

Now we have all the main parts we need to think about adding glass to them. I have made the frames and door quite thin to leave room for a backing piece of glass.

For perfection, in the backing piece you could make a recess to put the acetate, plastic, glass or whatever in. This might be necessary with actual glass though not so much with acetate.

If you make the backing piece for the frame or door 2mm thick you can draw a box slightly bigger than the glass and make it, say, 0.5mm high. Then align it to the vertical top of the frame or door. This will then print with a recess. Note the size of the recess - 2.5mm larger than the glass sizes listed in the previous section.

Fitting jig

To make glueing the whole together more accurate, I have screwed offcuts of wood to a board, the bottom and left is a perfect right angle, top and right are detachable pieces to position them wherever they are needed for each job. This makes sure pieces are flush, but also allows you to add a little pressure to slow set glues such as contact glue or epoxy.

I made a small error on this one as I didn't have my chopsaw available at the time. The corner isn't perfectly joined which could allow very small things to slip inside. It is best to be perfect with this.

Part 3 - the shop inner

Walls and shelving

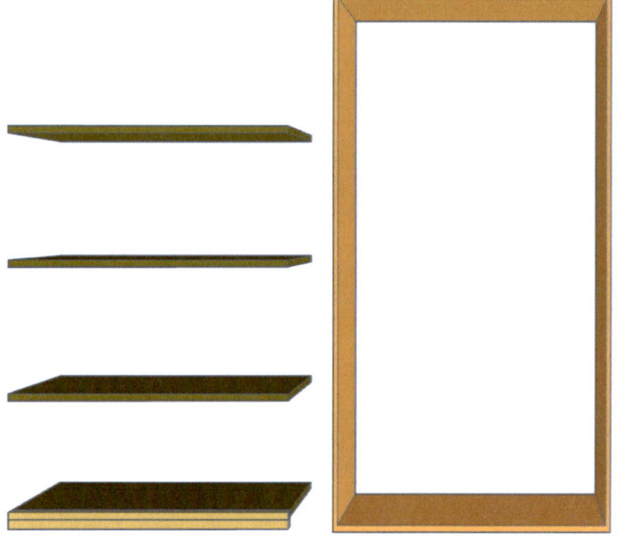

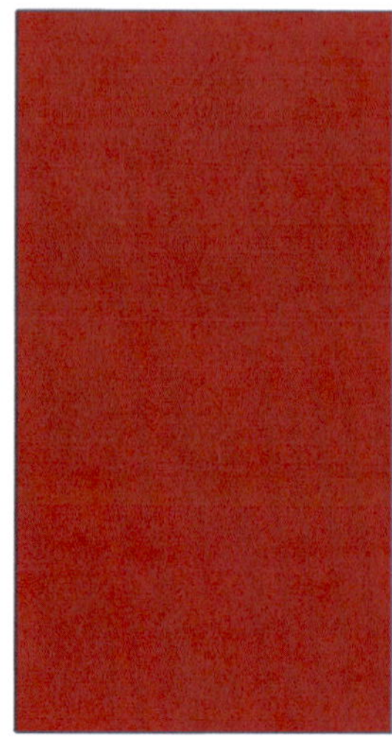

3 lower pieces and 3 upper pieces but serious issues with my Sidewinder printer soon stopped that, so after getting one of the 3 tops and bottoms I had to revert to printing the others as Ikea style flatpacks. Regardless of the method the result is the same and having the jointing may look more authentic.

Making a modular system of display areas in your shop.

Making floor to ceiling shelving units reduces the need to think about how to build walls and provides a solid place to attach other walls. The other walls then don't need to be printed, they can be cardboard or MDF for example.

I have copied the width of the shop table (146mm) to the shelving so it matches visually,

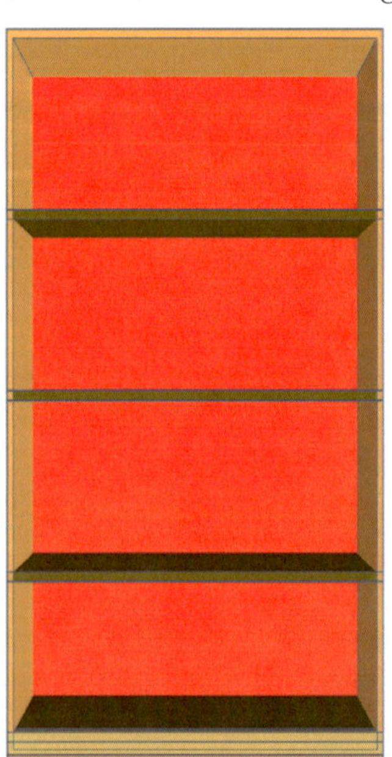

and put one set of floor to ceiling shelves at each side of the shop, then in the middle added a narrower shelf unit to fill the space. Alternatively you could put a door to the storeroom here. In this project I have put a door to the side where an understairs cupboard might be.

Initially I aimed to print these units as

The main units total 146mm wide by 175mm high, with a closed-in bottom kickplate forming a base and 3 shelves. The kickplate is 11m high and the first shelf is at 70mm to match the counter height of the display case and table. How you arrange the other shelves is personal taste but I arranged them equally inside the remaining space.

The shelves are 3mm thick and 52mm deep with a piece right across the back to close them in.

I copied all this over to the middle, narrow, shelving which is 81mm wide.

Finish

The surface appearance of a print can be changed in the slicer. By default it may have slight lines or a texture.

You can make it very smooth by "Ironing" the surface, this scrapes the head over the top surface until it is smooth, but takes extra time. **Ironing**: Print settings -> Infill -> Ironing

If you want a random, organic feel go for **Fuzzy**: Print settings -> Layers and perimeters -> fuzzy skin

Add on software lets you texture the surface in any style you choose, wood, stone, other natural textures.

Fake door

Of course there should be a door inside the shop to the under-stairs cupboard, storeroom etc. I copied the proportions for the panels to the body of the door to the surrounding frame from a photo of a Victorian era inner door.

Then I copied their midpoint from the existing front door to try to keep them in a similar era.

Make a simple box for the door, I went for 161mm by 69mm wide.

Make up recesses for the panels in your choice of era, I went for upper at 17mm by 75mm and lower at 17mm by 44mm from the photo I used.

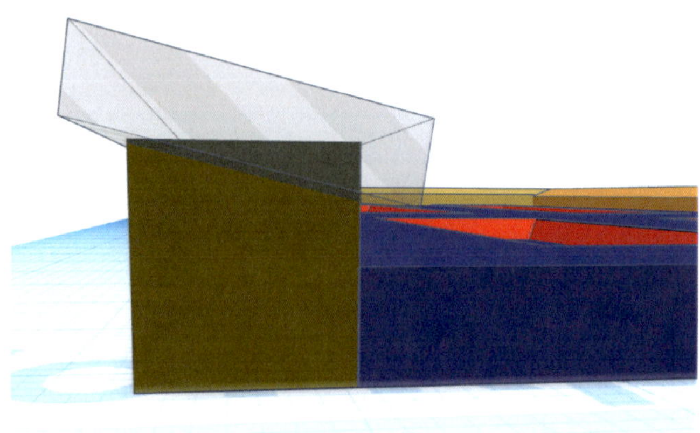

panel, then copy them to the lower panels and stretch the sides down.

I copied a door knob from the front door.

I made the door surround frame in 3 pieces to make it easier to chamfer them individually, this sizing is just done to eye, from the photo I got approx 7.5mm. A piece for the side mitred off at the top, mirror this piece to the other side and make a top piece mitred off at both sides.

I made the frame (e.g. 8mm) twice the height of the door (e.g. 4mm) and used the angled hole to split the difference at the inner edge (e.g. 6mm)

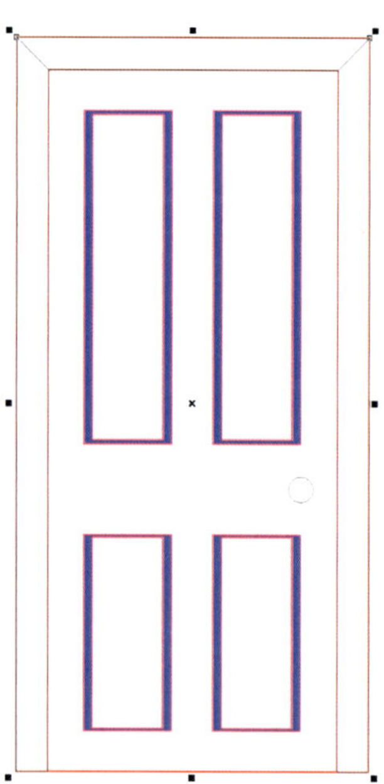

Using a midpoint gap of approx 21mm from the existing door and spacing everything equally this gave me a layout.

Next I mitred a 2mm cube at 45 degrees to give me beading for the panels. I stretched and copied it all around the panels. It's fairly quick if you do an upper one then copy it to the next

Glass display counter

Using the same concepts as the shop table, and the same dimensions, we will make an old fashioned display counter full of drawers that taper off towards the back. It will have 3 shelves of progressively shorter drawers.

The first diagram just copies the dimensions of the shop table (and later the shelving units).

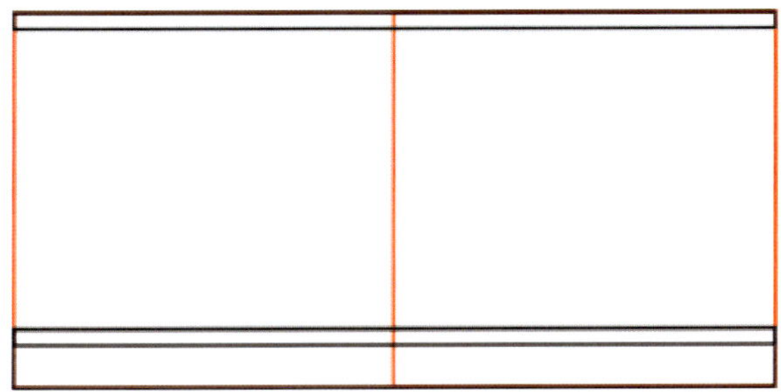

This is a little more complex than previous pieces as it needs a few diagrams to get the measurements right for the shelves to be almost invisible under the drawers and the angle to be right for the drawer progression up. On the plus side these diagrams can be exported as SVGs to make the components to print.

First I made a box to get the width right for the drawers. I have made full length shelves with a hidden reinforcement section in the centre with slots cut in it for the shelves. All this has to be measured before making a drawer so that 4 fit in the shell comfortably.

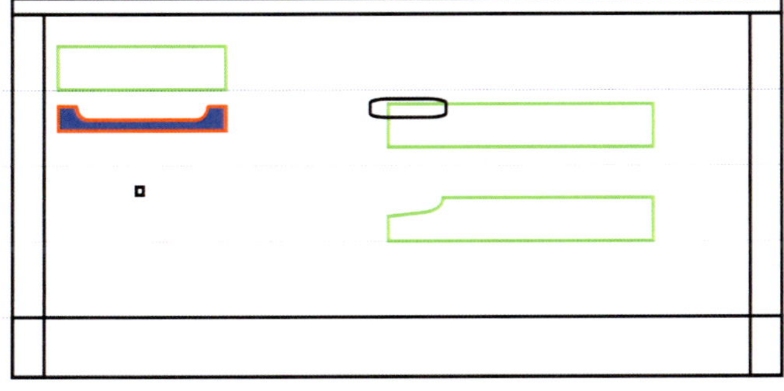

Then I notch the drawers at the front to display their contents better, and at the back for a handgrip to get them in and out.

I made crossbars with a cut in one corner to attach the glass (acetate) to and notched the side pieces to fit these crossbars flush.

Next I made the height of the shelves to fit in a kickboard at the bottom and have room for 4 drawers in total vertically accounting for these notches in the corner of the unit.

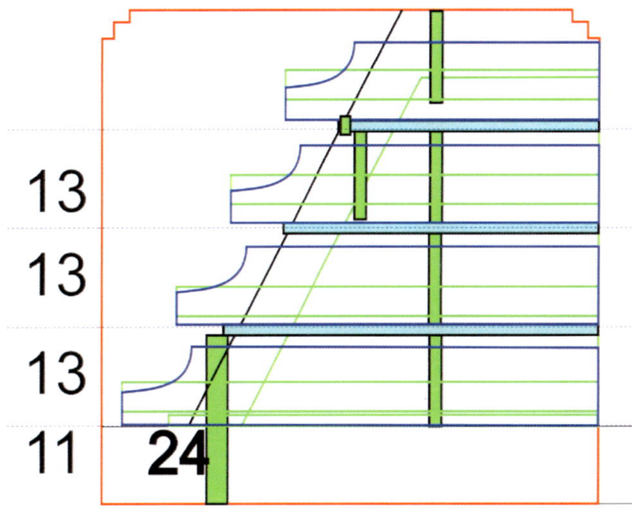

The green pieces in the diagram are just for measurement. I went for 11mm high for the bottom kickboard then 13mm spacing for the shelves above it.

Looking back, to make it perfect, the upward curve of the drawers should be the starting point of the next drawer up, but it's made now.

I drew a line at the front of the shelves diagonally to make the centre support, and cut notches into it for the shelves. The shelves then had a notch cut into them the size of the green box to make a strong inter-connection.

Finally this diagram also gives me the positioning for shelf supports on the side pieces.

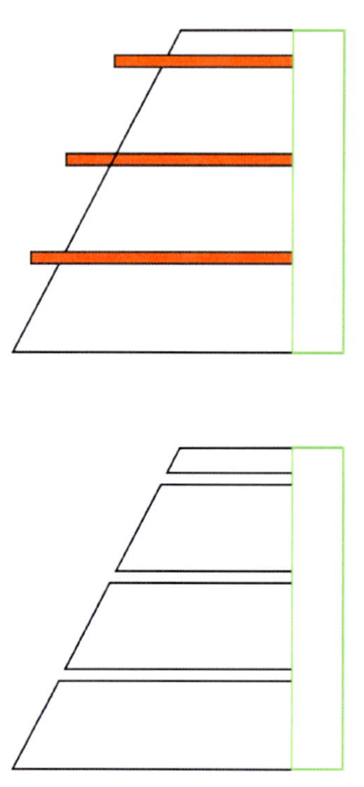

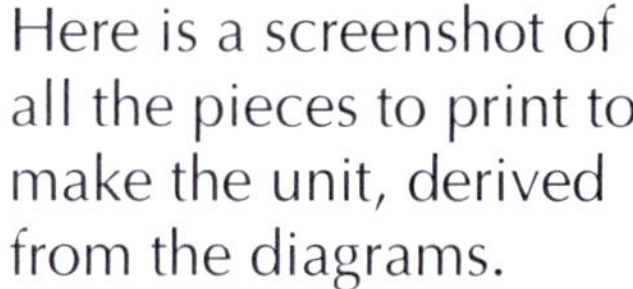

Here is a screenshot of all the pieces to print to make the unit, derived from the diagrams.

3 shelves, progressively smaller and notched at the back, centre, 1.7mm by 8mm. The sizes are 37mm by 142.5mm, 45mm by 142.5mm and 53.5mm by 142.5mm.

A centre support for the shelves. 52.4mm at the base, 26mm at the top and 49.4mm high, notches of 1.7mm full width except 8mm at the back, at positions 24mm, 37mm and 50mm.

2 Sides - notched and mirrored with shelf supports of 1.5mm at 21mm, 34mm and 47mm.

2 full width crosspieces for the top 146mm long, 4mm square with 2.5mm removed matching the notches on the sides, and 1 internal bar of 1.5mm by 142.5mm for the bottom to hold the glass.

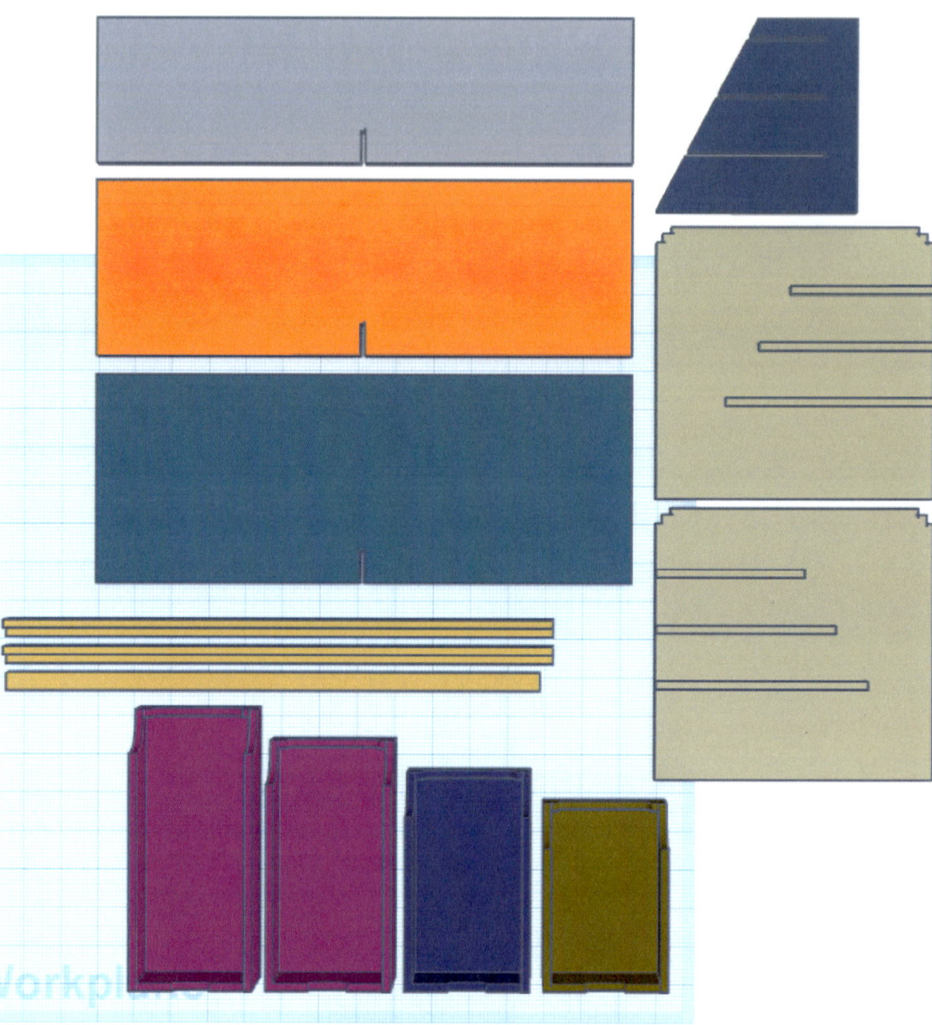

Finally 4 depths of drawer - 1 of each. Width 32mm, height 11.5mm and depths of 70mm, 62mm, 54mm and 46mm.

I gave this a test print in 24th scale which was quite quick and let me test the drawers sideways and vertically and I was happy with the result so I could print the full size unit with 16 drawers.

For the glass I have already cut 142mm by 68mm and 142mm by 56mm from acetate when I was set up to cut the window and door pieces.

Shelving insert drawers

Based on the smallest drawers from the glass counter project I made a stand to fit into the larger shelf units. The sides curve back from the front so as not to present an ugly double thickness panel, and 3 shelves are built into the print.

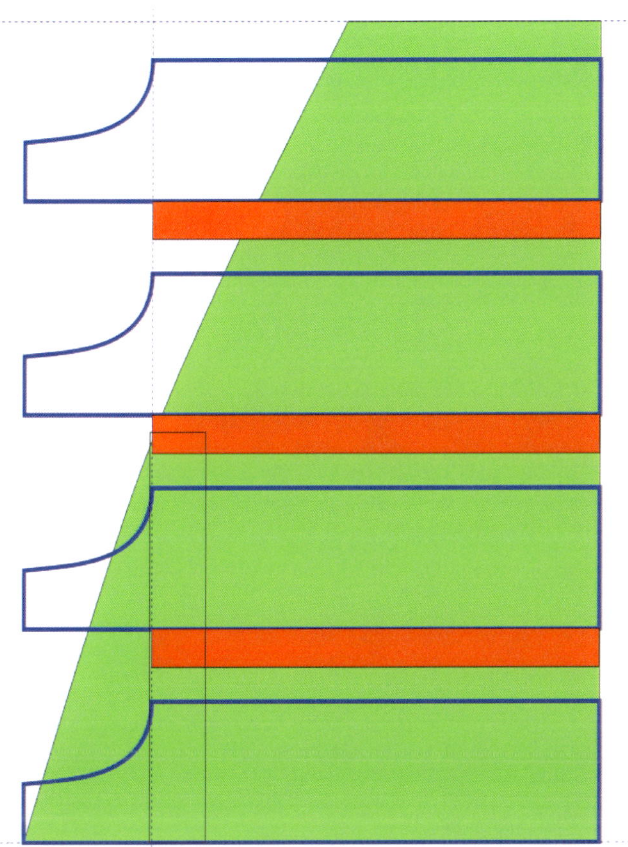

You can probably see that if the shelves are 3mm thick they have 3mm clearance between them and the top of the drawer (height 11.5mm).

The shelves are 33mm deep by 142mm wide and 3mm high - full width into the side pieces.

The curved side and centre pieces are 3mm also, 46mm at the base , 20mm at the top and 64mm high.

Group the unit and print it in one piece laid on its back.

Clothes rail and hanger

This is a simple tube project made of rotated cylinders, and a simple outline drawing for the hangers in an SVG package (e.g. Inkscape).

For the hanger rail I drew the proportions out using simple rectangles and 45 degree rotation in an SVG, then in Tinkercad I went for crossbars of 4.5mm diameter and length 120mm.

To make the side support I made cylinder joints to receive the crossbars with a 5mm hole, outer of 8mm and height of 4.5mm.

These go on either end of the side pole (copied from the crossbar and shortened) of 108mm.

Next I copied two poles 17mm, and 2 poles of 32mm rotated 45 degrees.

Push these together to form a structure and print 2 of these and 2 crossbars.

For this print I had to use Brim for the crossbars to keep them firmly on the bed, the sand it away afterwards. It is worth going for the smoothest print setting for this as it is a very simple item.

First of all in the SVG package I imported photos of coat hangers and drew something

that approximated them, I added more strengthening so I could print it in PLA, and cut away more of the hook so it could fit on a stronger, thicker crossbar rail.

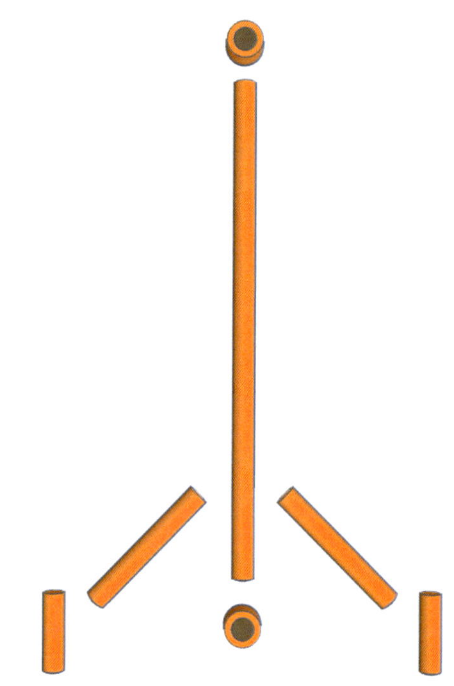

Next I exported the hanger and made it 1.33mm high which was enough to print with some strength without being too thick.

Hang the coat hangers on them and it is done.

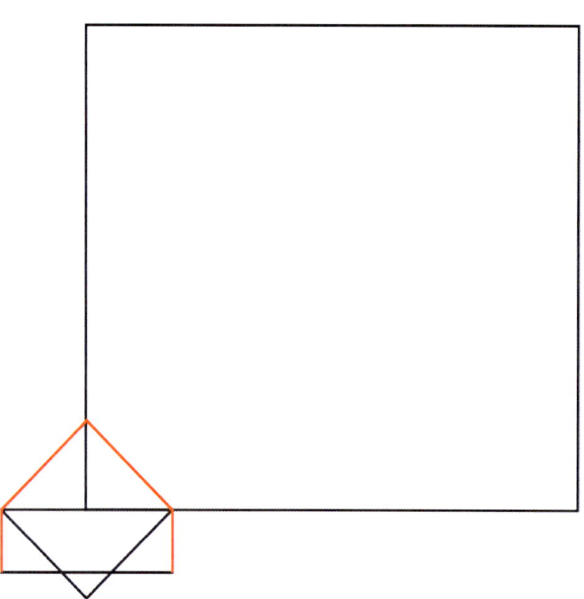

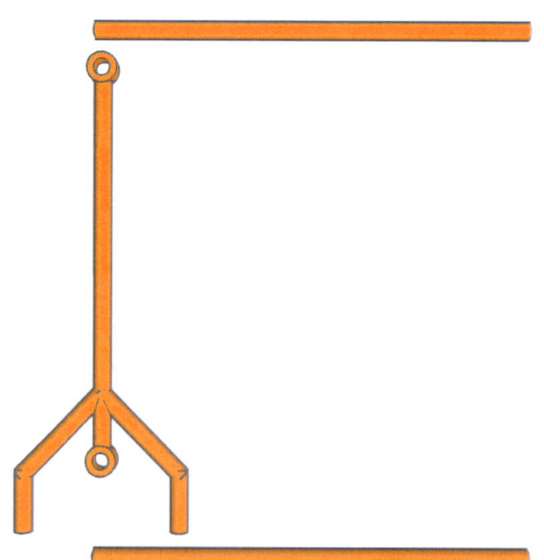

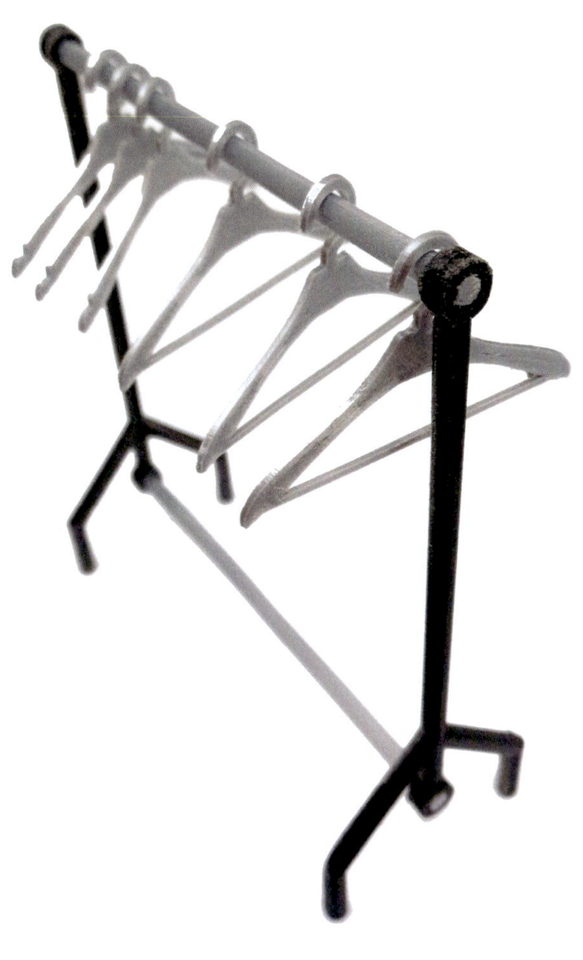

Ribbon display

Spools

The spools are made of a flattened cone, a cylinder and a hole. I have made a variety of heights and widths for different ribbons we had available in 12th scale but here is an example one.

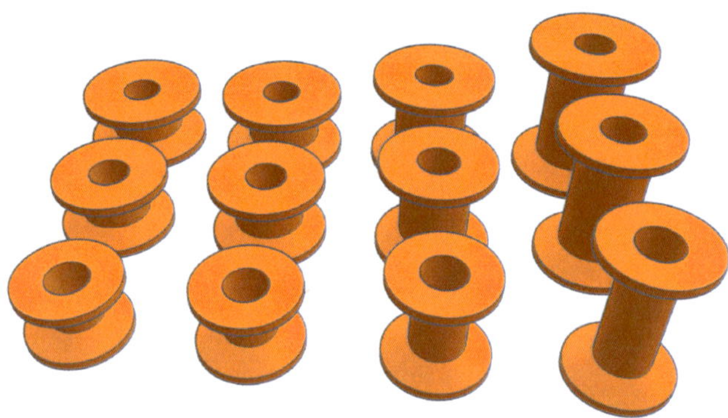

Make a cylinder of 15mm diameter and 1mm high. Add a cone on top of 15mm diameter and 3mm high. Choose your total height for the spool, taking 2mm off for each end will give you the width of fabric it will hold. I chose 8mm high giving approx 4mm of ribbon.

Add a cylinder of 8.5mm diameter and 8mm high, and a cylinder hole the same height but only 5.5mm diameter.

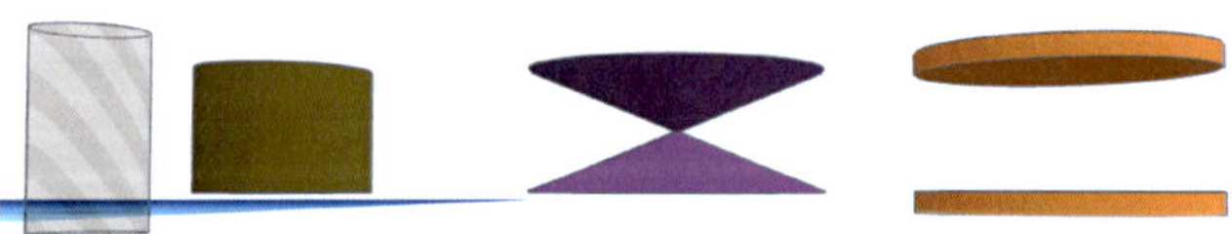

Centre them all and group to make a spool you can print in one piece. If you want perfection you could do away with the cones and print in 2 pieces then glue, but this is pretty near once the fabric is in place.

Display unit

The display is a curved piece drawn as an SVG then holes added to slot dowels through (made from cylinders). The two ends are 61.5mm high by 45mm wide and 3mm thickness.

The flat base is 74mm by 45mm and 3mm high, the back is 61.5mm by 74mm and 3mm thick. These measurements are the inner space of the middle, narrow, shelves minus a small amount to ease fitting.

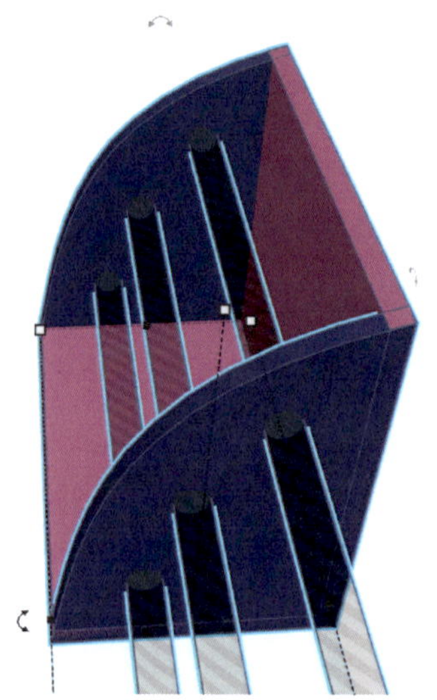

The dowels are 4.5mm diameter by 74mm long, and the holes in the side are 5mm diameter.

The dowel cylinders have Brim added to make them easier to print, then they need to be sanded clean. I have made the display snugly fit into the shelving and this holds the dowels in place too. Load them with spools of ribbon and insert the whole into the shelves.

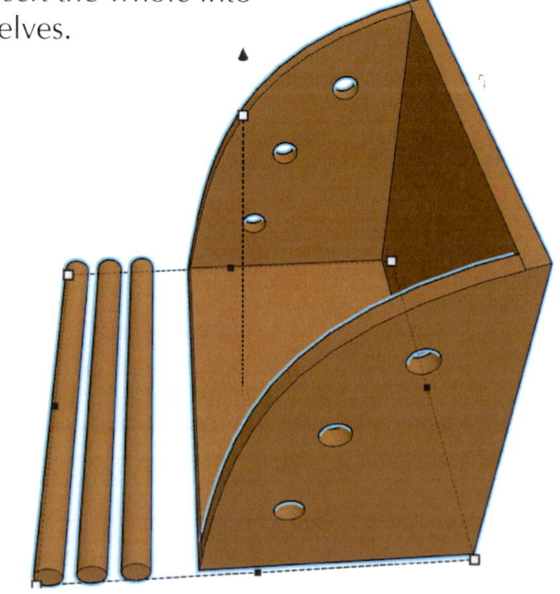

Vases 2 - using the lathe tool

Some software has a "lathe" tool that takes a 2D profile shape and spins it into a 3D shape. The software I used in the 90s made this so simple. However it seems the software that does this today is much more difficult to learn, so here I have a simple method that works on anything including Tinkercad. First note that Tinkercad has an addon called SVG Revolver that may work for you on some designs, though I haven't found it willing to accept shapes from the SVG software I use, and it also has limitations. This manual method lets you make a smooth-as-silk model and isn't limited to the number of steps in the rotation, you can go round more than 360 degrees easily adding in half degree steps or less if you want.

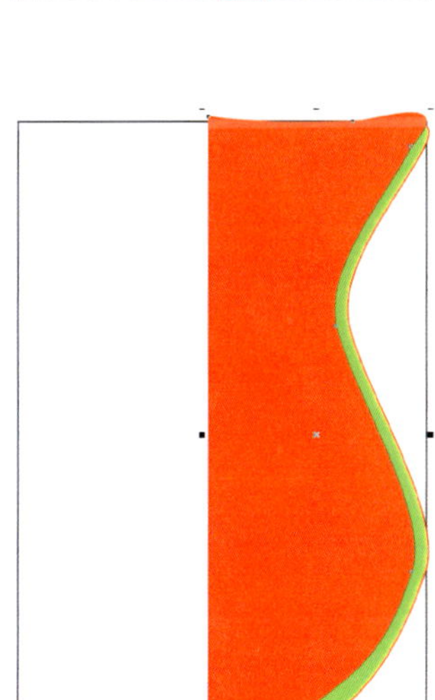

First draw only one side of the shape you want. I suggest using a box to note the exact centre as symmetry is very important in this process. Draw it at the size you want it to print.

Next decide how thick you want the outer edge of the shape and move a copy of the image out to there. You may also need to make minor adjustments along the shape. Using filament remember you shouldn't go too near the 0.4mm standard nozzle output or parts may break and curves upwards may become too difficult to print. With resin you can be finer if

need be. I suggest it is only important at the visible top edge and further down inside the vase keep some distance for thickness and strength.

In this example I made the base a bit too small to be sure to print it well - it failed at the first attempt - but you can add a short cylinder at the base later to fix this kind of error.

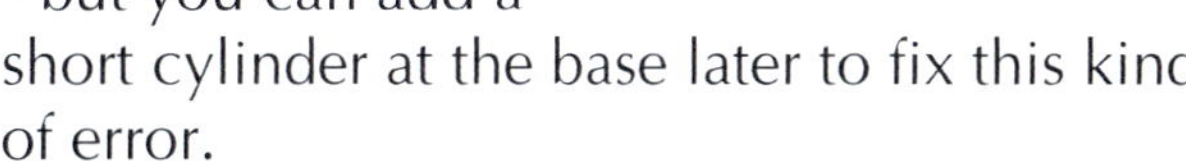

You can use Path -> Difference to leave a line of the curve you want to print.

Now flip it horizontally, then weld the two pieces together. Export to SVG.

Import into your 3D software at only 0.1mm high. I am going to rotate this 180 steps. If

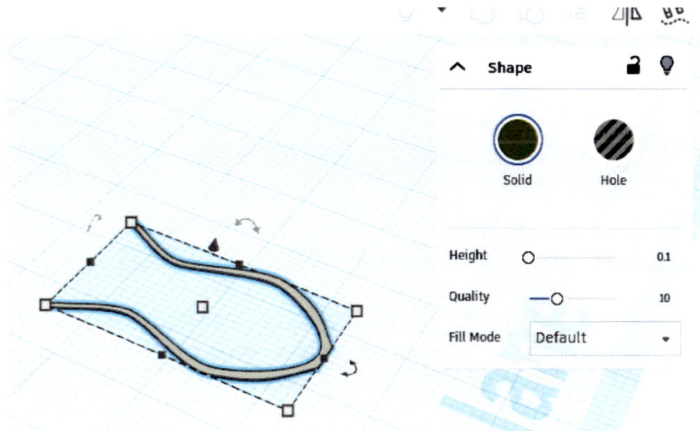

you are going for super precision maybe adjust the height down a little further. You need it thick enough to cover the whole shape but not so thick it juts out

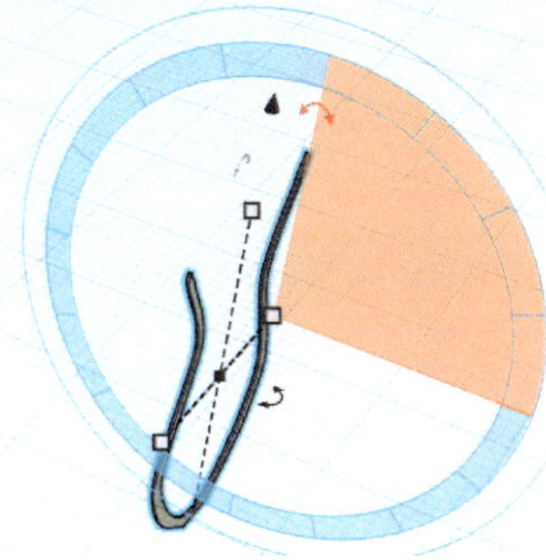

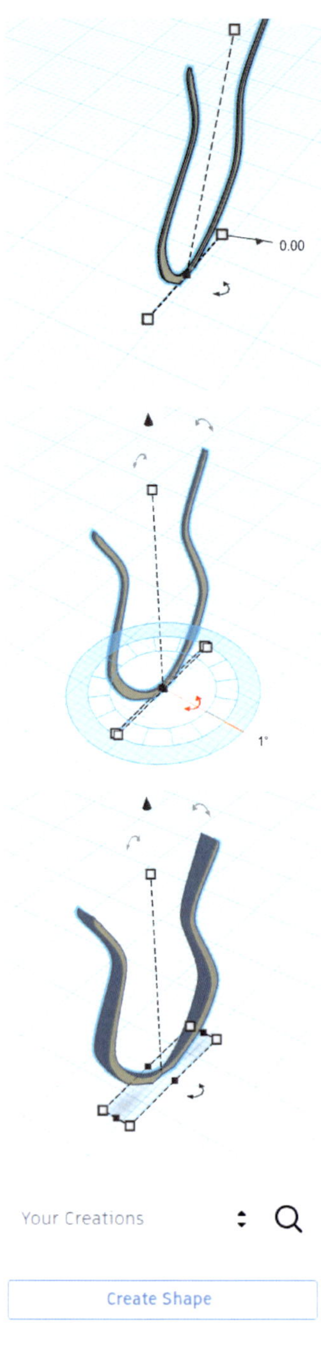

leaving visible lines all around.

Rotate the piece upright, then move it up to the baseline if it has sunk below the ground level.

Next we will use the duplicate and repeat feature. This copies an object and applies an action as it goes, so I will copy, apply 1 degree of rotation, then repeat until I have 6 pieces.

I want 180 pieces, but that is a time consuming way to go about it. Also it will bite you later as the object will become huge and slow in memory and probably not export. So my next step was to use the simplifying feature, which takes the 6 pieces I have made so far and turns them into one simple piece - select them, say Create Shape and give it a name e.g. Vase Slices. I then repeat the process with the simple piece but rotating 6 degrees each 6 times to make up a 5th of the vase creating it as "Vase Chunk". Then copy and rotate 36 degrees until the vase is complete and Create Shape, name it Vase Complete. You can now drop this shape into designs as if it was a primitive such as a cube or sphere.

Notice in the first slicer diagram that somewhere along the process I missed a piece! However, I can just rotate yet another copy and it is fixed.

You don't have to make solid shapes with this technique either, you can make interesting latticework or other patterns with it.

In the final picture you can see a line right up the vase, this is caused by the printer stopping in between each layer as it moves up. To remove this from your final print you need to select "spiral vase mode" in the slicer.

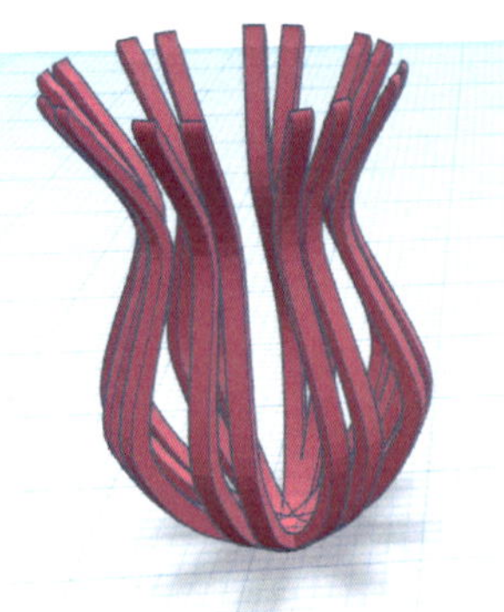

Shop cash register

The keys on this look complex and seem to hang in space, but there's a secret to keeping them there! It is a little fiddly but not technically difficult. Patience and taking copies at each stage may save you headaches.

For this one I looked at old cash registers and decided on a rough shape. I then planned it in 3 planes: front, side and top. The green view is side (the drawer is in yellow), the red view is front, the blue is top.

This let me get proportions and positions for the key holes, drawer and price display. The black quarter circle is for the key mechanism.

I am going to print this cash register in 5 main parts. First the upper body and keys of the cash register, then the bottom that holds the drawer, in silver. It is easier to split them and glue later as there would be a lot of bridging for the printer otherwise and a chance it could sag and the drawer not fit.

I will have a solid black base to sit these two on, and a black trim just under the keys. Finally is the drawer with sections for notes and change.

The first fiddly bit is the key mechanism. The lever is a quarter of an 8mm cylinder at 0.8mm thick. The keytop is a cylinder of 3.5mm and 0.8mm high. Then there is an inverted cone of 3.5mm diameter and 1.5mm high to support it in place.

Copy the key to make one for each hole and align them in front of the holes in the diagram but not touching, leaving enough space for the keys to be out into thin air at the front of the till. Align the top of the semi circle to the top of the hole.

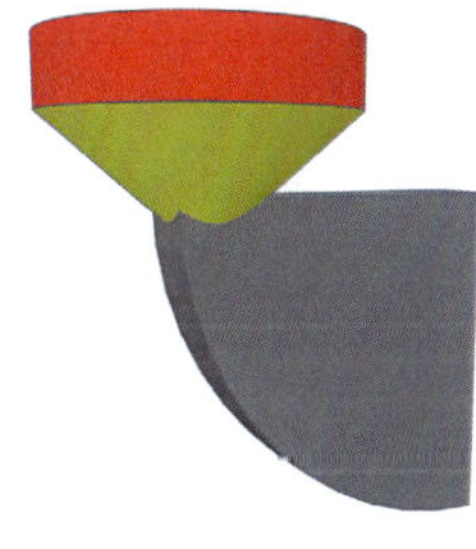

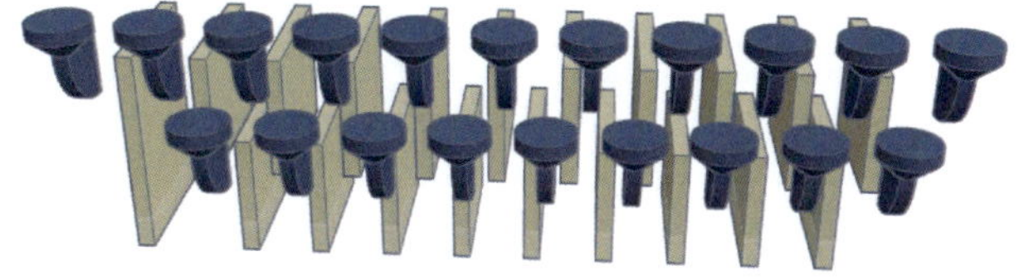

Next add in support bars to hold the keys in place. Group the keys and support bars.

Bring in the main body SVG and make it the same width as the base support. Carefully move the holes and key group around - front to

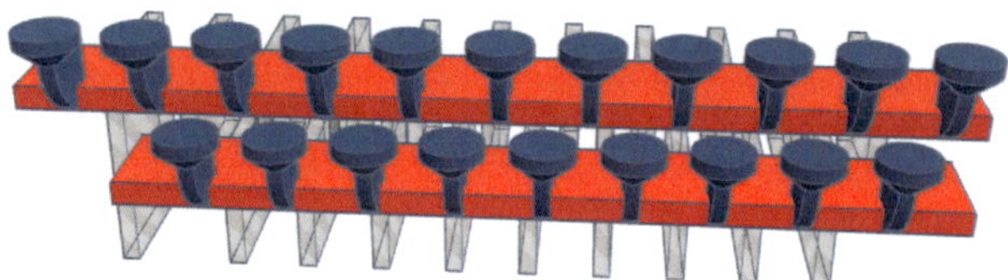

back, up and down - until you are happy with the look and alignment. This is worth being patient and working closely from different angles.

Add the display holes at the top, both sides. Later you can print something to go in here.

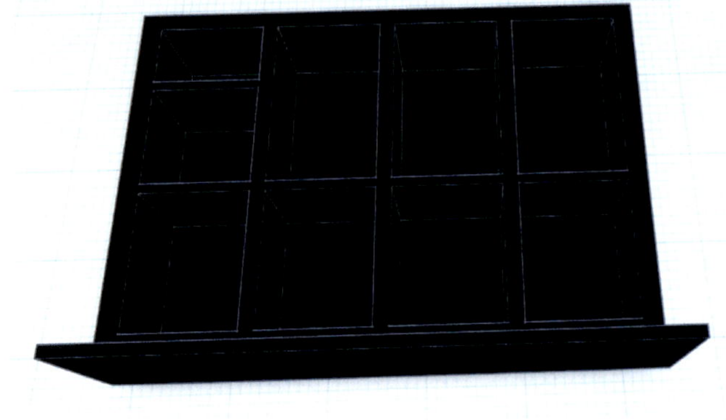

You can paint the keys red for the top left, the others black. Also print an amount in white on black for the price display.

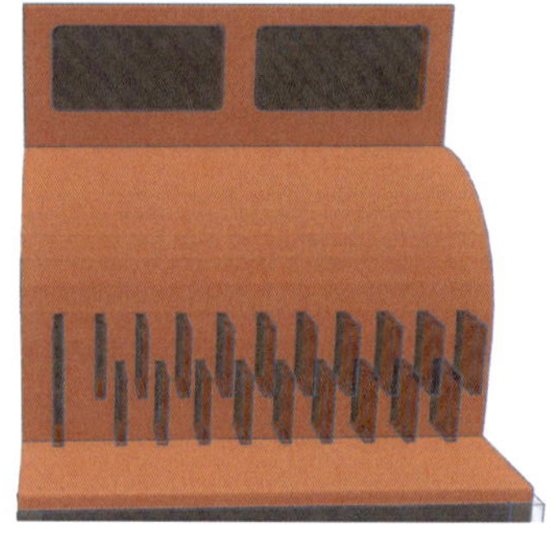

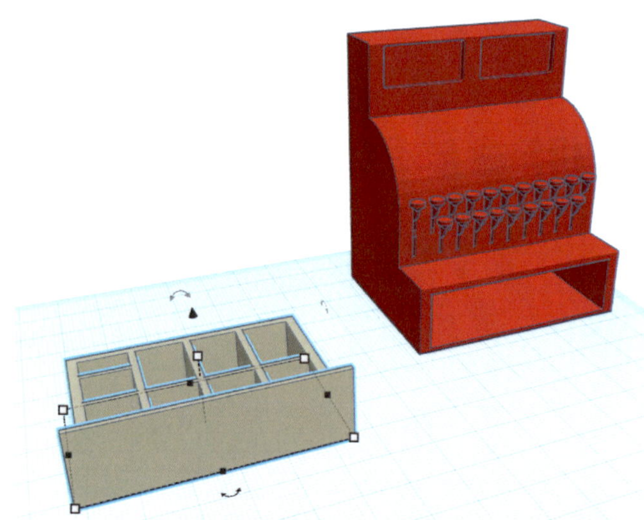

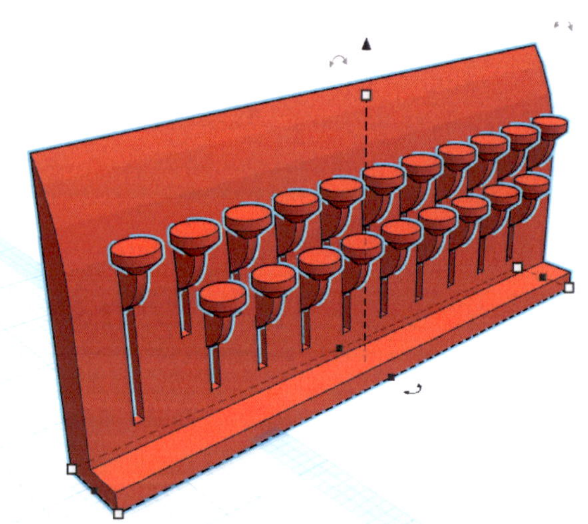

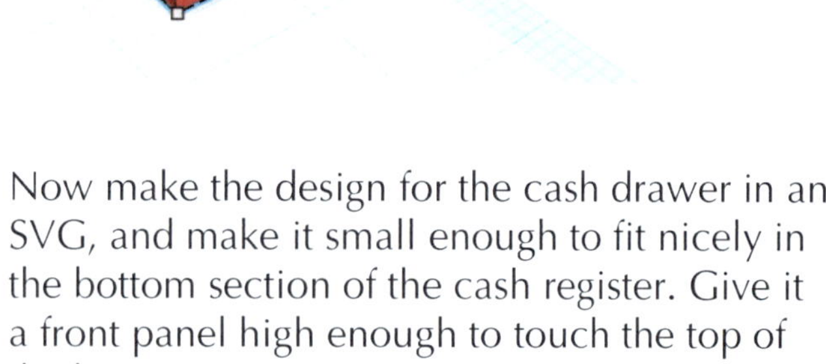

Now make the design for the cash drawer in an SVG, and make it small enough to fit nicely in the bottom section of the cash register. Give it a front panel high enough to touch the top of the bottom section.

Add the base and the trim section under the keys to finish.

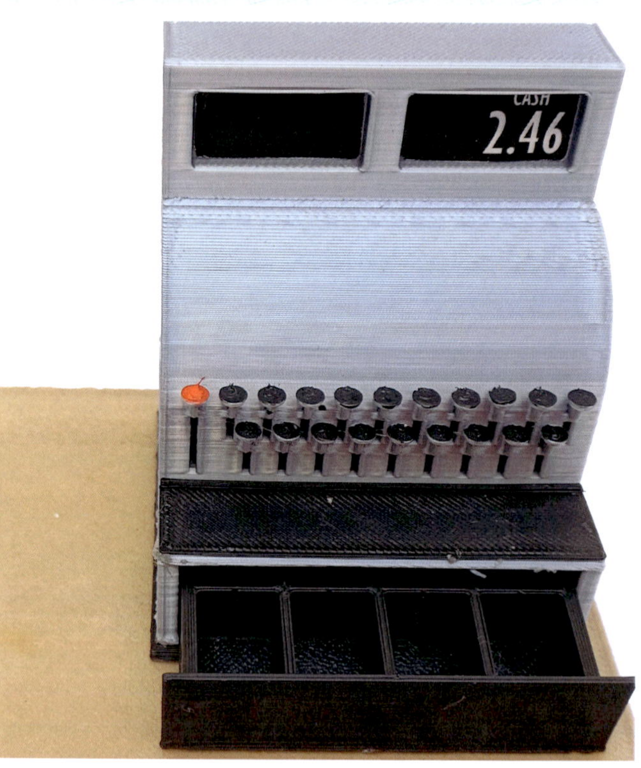

Part 4 - 3D Printing Basics

- Tinkercad 3D modelling

- Inkscape 2D creating SVGs

- Prusaslicer - turning models into prints

- Design basics

- Questions and answers

- Glossary

3D Printing Basics

Printer parts

The machine is normally one of two types - filament or resin. Here is a quick guide to the important pieces.

A filament printer will have:

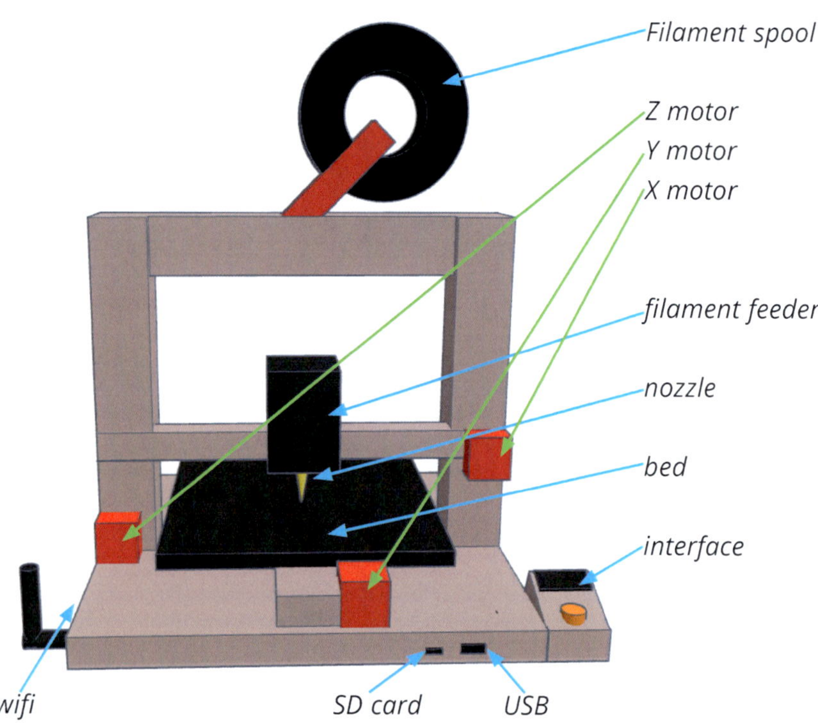

Controller / circuit board - where the movement instructions are processed (internal)

X/Y/Z motors - to move the bed (normally Y - front to back) and extruder (normally X - side to side, Z - up and down)

USB / SD / Wifi - to transfer printer files

Interface - a dial and buttons or a touchscreen to control the printer

Filament - the material (e.g. PLA) used to make the prints

There will also be **fans** to cool electronics and filament, and **structures** to hold all these parts together.

Extruder - the hot part of the machine that moves and ejects the filament

Nozzle - the part the filament is squeezed out of, it can come in varying widths.

Bed - the base for printing onto, normally heated

Filament feeder - where the spool of filament is pulled into the extruder

A resin printer will have:

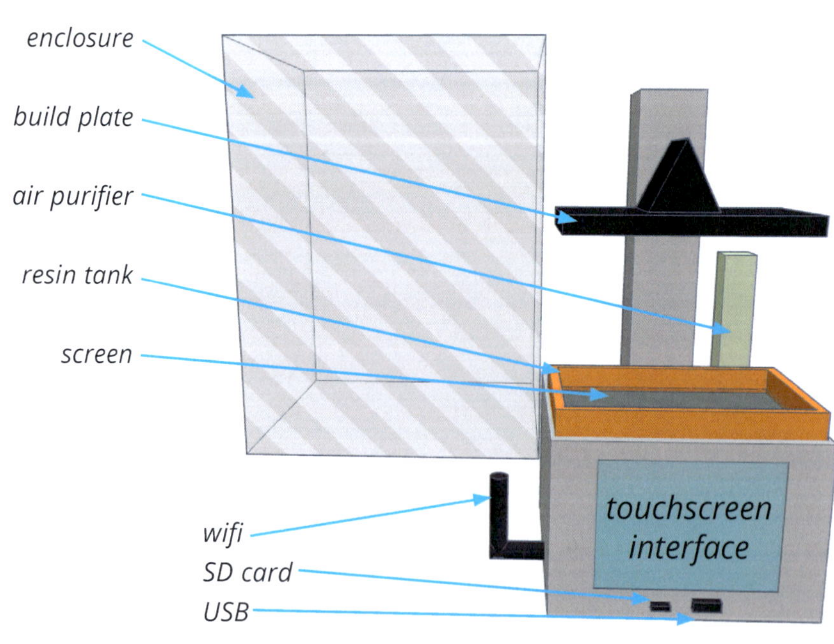

Screen - to project a slice of the image onto the liquid

Bed or build plate - to hold the print, which is lifted vertically away from the screen at each stage

Tank - to hold resin near the screen

Film - transparent plastic between the resin tank and the screen

Controller / circuit board - where the movement instructions are processed (internal)

Z motor - to move the build plate up and down (possibly internal)

USB / SD / Wifi - to transfer printer files

Interface - a dial and buttons or a touchscreen to control the printer

Resin - the light sensitive liquid

Curer (optional extra) - to set the model firmly after print

Washer (optional extra) - to clean off excessive resin

A resin printer may also have **air filters** as the fumes from them can be toxic depending on the resin used, and will be in an **enclosure** that blocks light.

Modelling

You need to download or design something to print (see design section). There are free models and free design software out there already.

Slicing

Starting with a free slicer like Cura or PrusaSlicer worked well for me and I haven't felt the need to upgrade to anything else yet.

There is always setup time for these programs, where you tell it the machine you have and the limits it has to print. For instance I have a Geeetech A10 with a 22cmx22cm quoted print

bed area and a 26cm maximum height. The height is apparently optimistic and prone to crunching the cables so I never go past 22cm.

Next you select the filament you will use (PLA, PLA+, PETG etc.), normally there is a handy list of presets to make this just a tick from a selection, but if something is a little different it is possible to tweak this too.

Finally, the fine tuning of settings. I have made a lot of changes after finding various small issues like stringing issues (lots of messy strings on your print) and bridging issues (printing over holes). See later for more in depth explanations on these two.

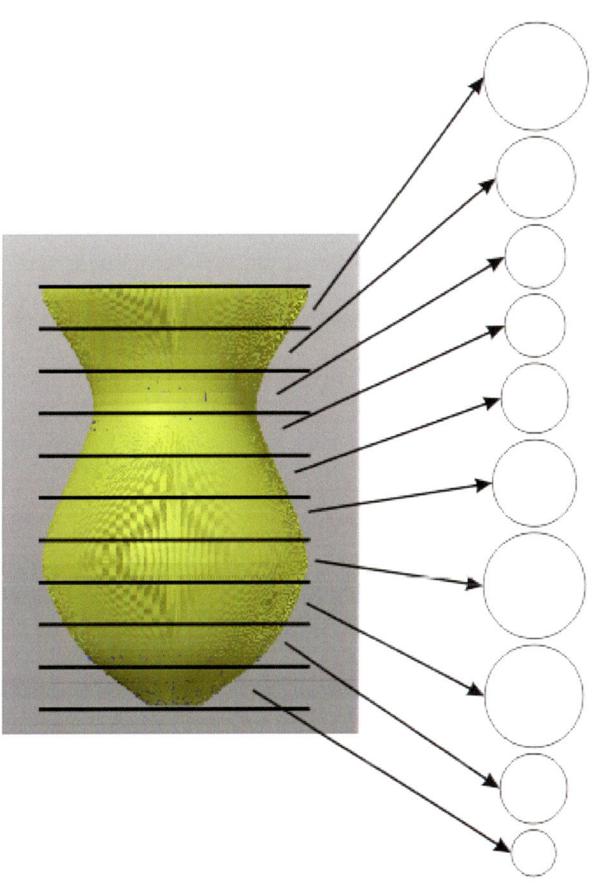

What is slicing? The software takes your 3D model and cuts it into sections to print it, in the simplified vase example it is turned into a succession of rings. Starting at the bottom the printer will take each ring and print it then move up a step at a time. In reality the printer will have many thousands of these steps at 0.1mm, which is the finest for filament, potentially even finer for resin.

It relies on the heat to glue each band together, then quickly cool to set as a solid object. However this can leave a line as it stops at each section, instead you can activate Spiral Vase mode to keep it moving around and up at all times.

Bed levelling

You must do this first and check it regularly on some printers. It is normally simple and sometimes automated. My printer is half automated - it moves to the corners but I adjust a knob manually so the head just scrapes a sheet of paper pulled over the bed. After setting all 4 corners do it again as there is often a change that affects the first corner.

You can get free test prints that will tell you if you are good to go such as **www.thingiverse.com/thing:1557995** *"Bed Level Calibration by zumili"*

Test print

When you unpack and set up the machine make sure to print the supplied test print. Of 3 machines, the 2 filament ones never made it to the end and I had to tune them in some way. The resin one printed it perfectly but it

was absolutely welded to the bed so I knew that future prints needed a "raft" if they were delicate.

Resin printing extra considerations

Resin printers submerge a platform in liquid as it prints, there is always a pool to be accounted for as you form a shape. Some designs can be hollowed out to save resin and this can be done automatically. You will probably need to add drainage holes too or raw resin will be trapped inside the print, again this would be costly. Remember the print is upside down and to drain towards the top. Experienced users recommend printing on a raft with supports, sometimes with the object at the best angle to accommodate them.

Filament infill

With filament it is normally pretty simple, you can just reduce the amount of Infill to suit, 10-20% on a large object often still leaves enough strength to be fine. Smaller pieces that may have some strain on them can go between 50% up to 100% for very small pieces.

I did have issues with large flat surfaces and 10% infill, though this might have been an issue with my printer which failed shortly afterwards.

3D Design

There are many programs available to make a design, a confusing array in fact. I first used one in the 1990s that soon became obsolete, then I found Blender - and stayed away from 3D for years afterwards because of its infinite possibilities bringing complexity!

You may find that to start with and to build confidence a simple web based program like Tinkercad or Sketchup lets you create all but the most complex elements. There are various primitives available such as spheres, cylinders and blocks but you can also easily bring in a 2D SVG file you designed elsewhere, (e.g. Inkscape, Coreldraw or Illustrator). This can be given depth, rotated in any plane, and pieces added or cut out at any angle. I found it plenty good enough for moving designs originally designed for Cricut into 3D print.

Tinkercad

Viewpoints and zooming, moving

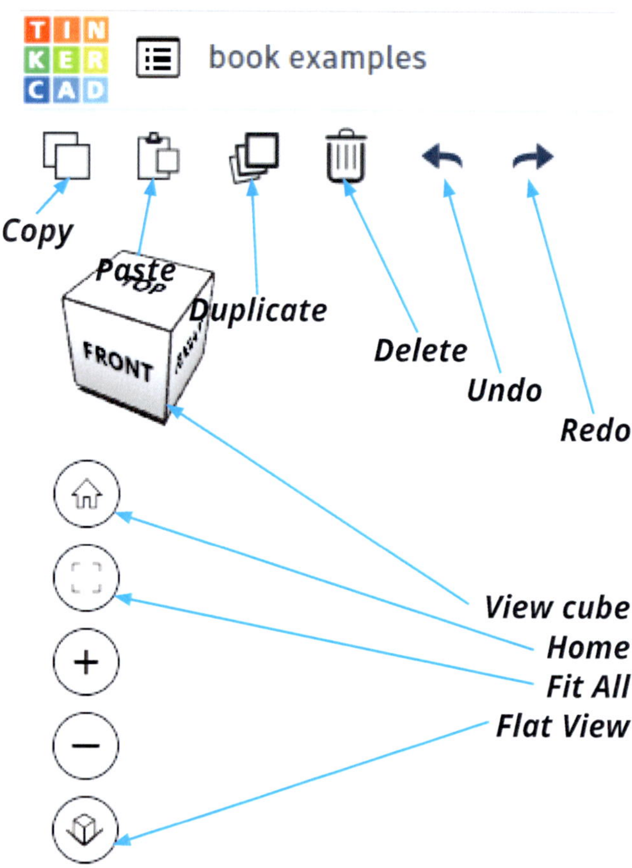

At the top next to the Tinkercad logo is the title of the current design (click to change) and the button to access your designs library.

Next is the icon panel with copy, paste, duplicate, delete, undo and redo.

Then is the view cube that you can move around with the mouse to adjust.

Under this is the home button to take you back to a view of the workspace.

Then the "corner box" (Fita ll in view) icon underneath will zoom you to a view of the selected object(s).

+ and - are zoom controls if you aren't using a mouse. Using the mouse wheel lets you zoom in and out.

Finally the Flat View icon will remove all the 3D representation which can get in th way of accurate positioning.

Holding the mouse wheel down lets you move the viewpoint across the workspace.

Practise these before doing an actual project to save yourself frustration.

Primitives

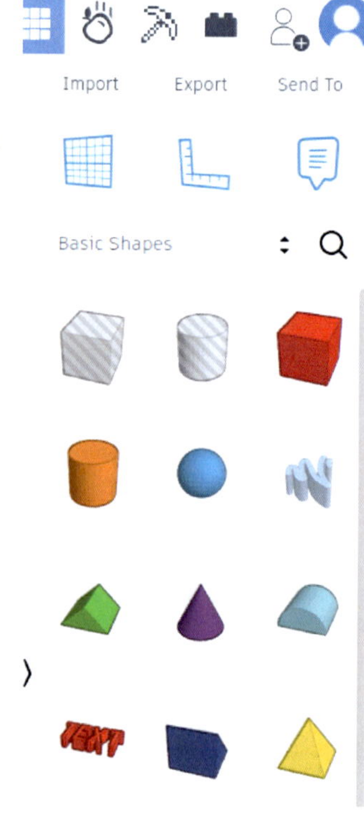

Primitives are simple cubes, cones, spheres etc. that you can use as building blocks in a design. In this book we will mostly stay on the Basic Shapes section as the others within it are generic clipart and specialist options.

Click and drag one of these shapes onto the workspace and then you can alter the parameters from it's menu. An example is Cylinder which

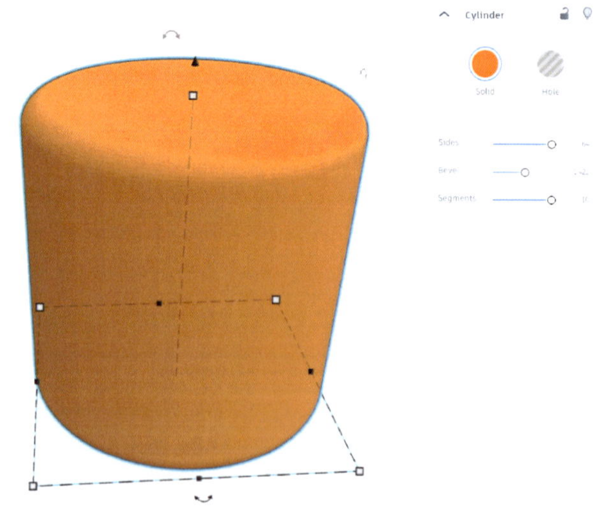

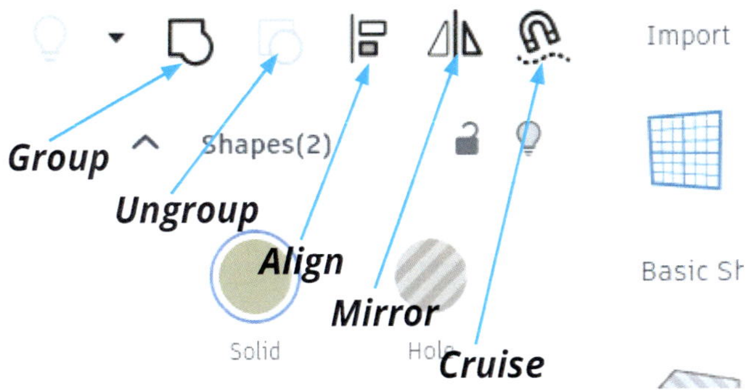

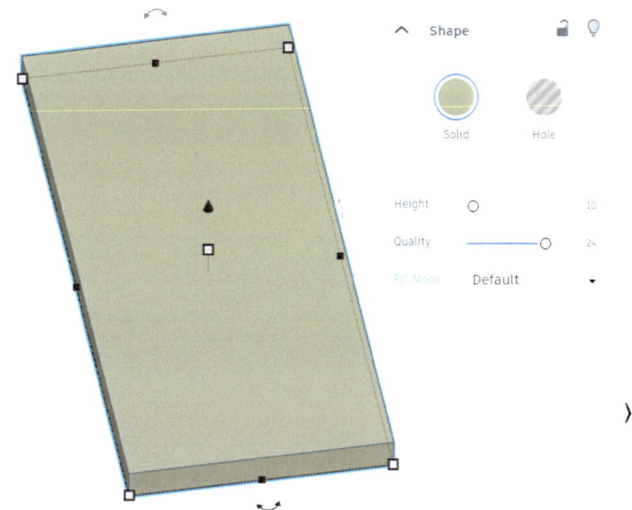

as well as adjusting diameter and height it has smoothing options. Bevel curves the hard edge down into the body of the cylinder. Sides smoothes the circular aspect of the cylinder, by default it is only 20 and can seem blocky if used large. Finally Segments smoothes the transition into the bevel.

Grouping, aligning, flipping

You can combine objects together which is useful for moving connected objects around, also at this point holes will be cut from other objects to leave the final shape.

You can align objects to each other at the edges or centre of any plane.

You can mirror an object in any plane too, e.g. flip it back to front or upside down without rotation.

Finally is Cruise which is a kind of Snap-to mode for assembling objects.

Import SVG

At the top right click Import, then find the SVG you want. It will be brought into the same place on the workboard as it is on your SVG design, allowing you to accurately overlay elements in multiple SVGs.

By default it is 10mm / 1cm high. If you click the white square in the centre you can then click the measurement 10.00 and type into it directly, in this case we want 2.5

Nudging

A simple concept - moving an object with the cursor keys. You can change how far it moves each time to suit, from 0.1mm to 5mm via the Snap Grid dropdown at the bottom right of the workspace.

Height

The small black cone in the centre adjusts height from the baseline, use Snap Grid bottom right to control how finely you can move it (if you are lining up a feature by eye) or type directly into the box.

Scaling

The white outer squares give you the dimensions of the sides and top, again if they need changing you can type directly into them.

Rotate

There are 3 sets of double ended arrows when you click on an object, each offers rotation in a different plane. These can operate in 1 degree jumps at the outer ring or 22.5 degrees in the inner ring, where it is easier to get a 45 or 90 degree turn.

Duplicate with operation

Using the Duplicate & Repeat feature you can make a copy, it will be exactly in the place of the first one. Then you can do a move, scale, rotate etc. and duplicate again and it will repeat the operations on each duplicate with precision. We use this in the vase project.

Create shape

Sometimes there isn't a primitive for what you want and you can model it yourself. You can cobble together whatever you need from primitives and SVGs until it is right, then go to Your Creations -> Create Shape. The shape will be welded and simplified and stored as a new primitive. Again we use it in the vase project to turn 180 shapes into one, simple, whole.

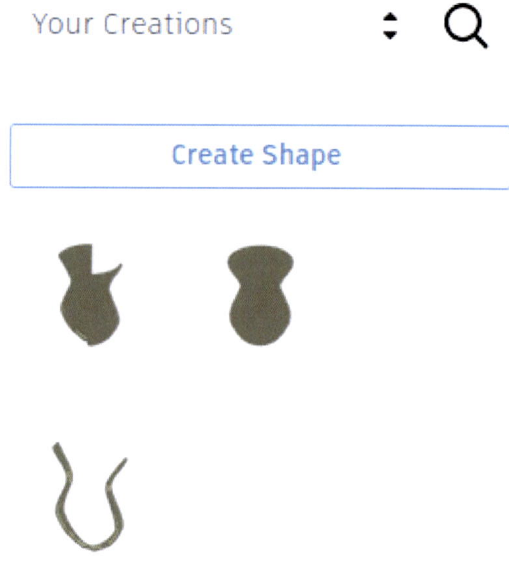

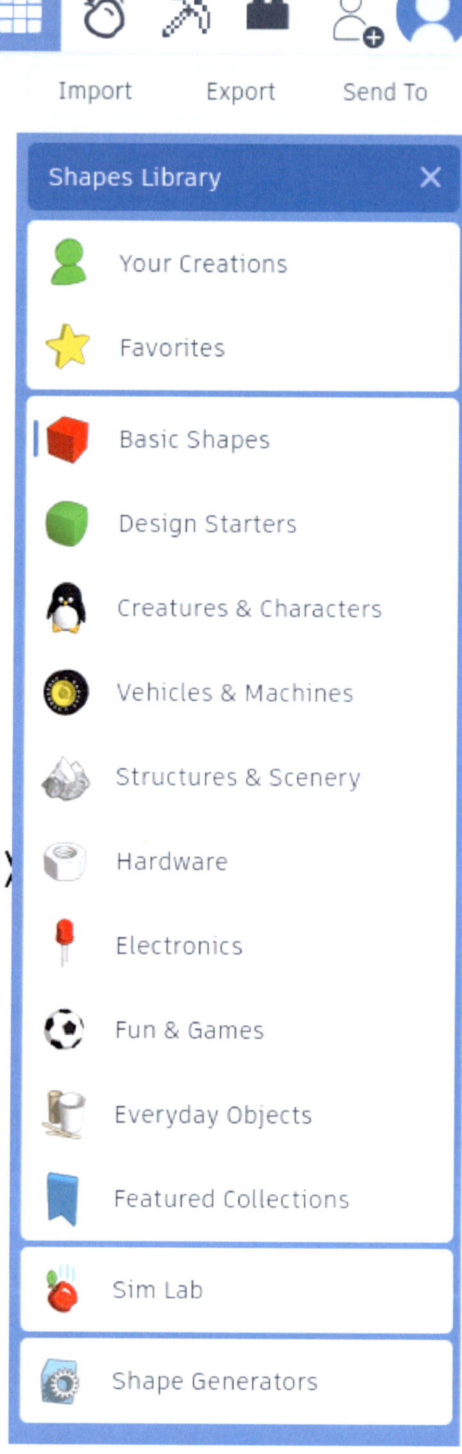

Scaling

If you want to make a version for 24th scale the easiest way is to select it all, hold shift, click on the corner white square and type 50% in one of the measurements. All sides will then scale at once.

The official tutorials are at
www.tinkercad.com/learn

Inkscape

Curves, paths, nodes

For the vase project we need to draw a curved shape that is perfectly symmetrical. Straight lines are easy but how to turn them into curves?

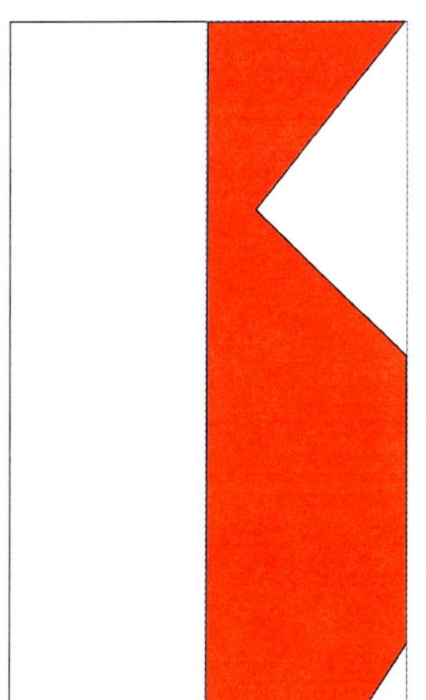

First I draw a box the total width and height I want, then another box exactly 50% of the width.

With the line tool draw the vague shape you want.

To add curves click on your object, use the Node Tool, click where you want the curve and on the top bar click "Add curve

handles". This will now give you two circles either side of the point that you can move freely until it turns into the shape you want.

Nudging

Moving an object with the cursor keys - you can choose how much from Preferences -> Behaviour -> Steps

Snapping

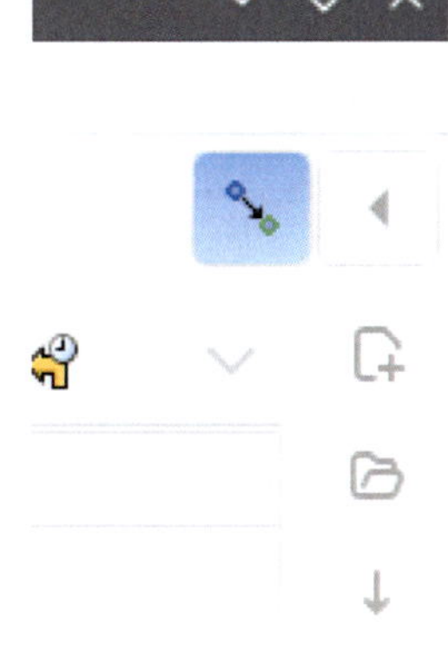

This is a useful assistant in making things fit perfectly. You can move an object to be exactly at the join of another. You have to toggle this on or off via the top control bar, hover over the icons until you find Snapping.

Group

A group is a collection of objects kept together, they move together and scale together. This helps keep relative positions intact, click Object -> Group.

Weld / union, trim, intersect etc.

Welding, or Union in Inkscape, is joining objects into one. Different to grouping, in that if you take a group to a cutting machine it will cut around all the outlines of the object, with Union there will only be the outline of the whole.

Select all the items, check they are ungrouped, and click Path -> Union. (There is an example later under Duplicate.)

To cut a shape out of another shape use Path -> Difference.

To get only the parts that cross each other use Path -> Intersection.

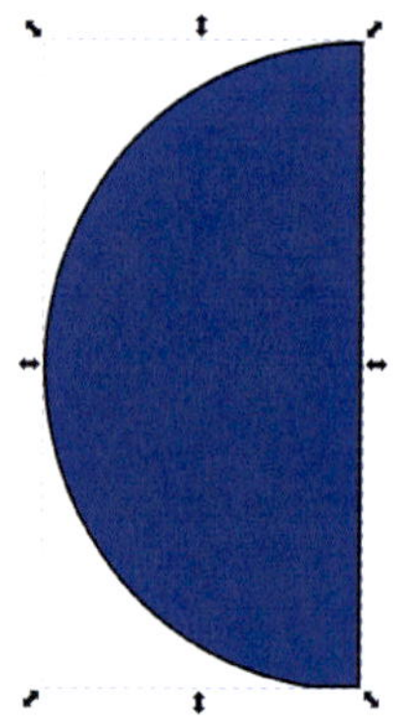

Some of this might not seem useful at the moment but quickly practise them as they are fundamental building blocks in design.

Closing paths

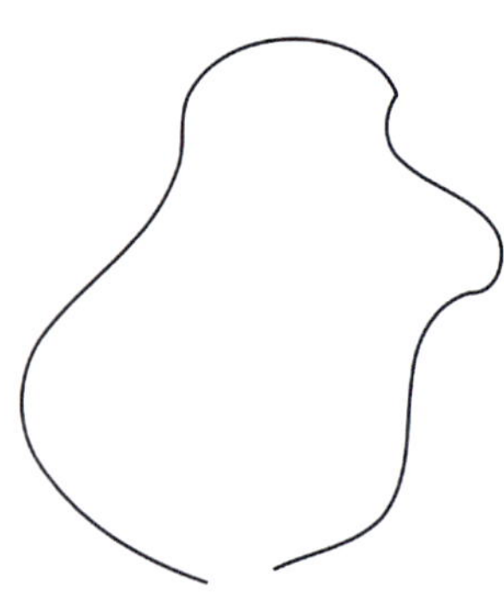

If you draw a shape and it doesn't quite meet up you won't be able to fill it or use it in Tinkercad. Select it and click Path -> Union.

It is auto-completed and can be filled and used.

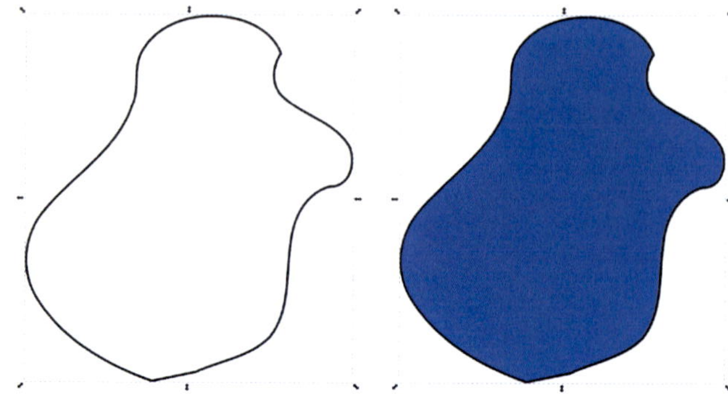

Tracing

Sometimes designs are so complex you will find it easier to sketch it freehand than make it onscreen using the curve tools. That's OK as most software has a trace option that will do its best to draw the outlines for you. You can then fine tune them until it is exactly what you want.

File -> Import and choose your image.

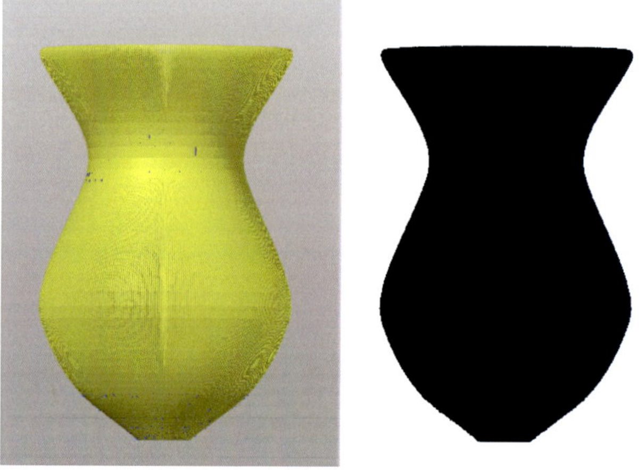

For this example due to copyright reasons I have chosen one of the self-generated pictures from the book, then tweaked it with a filter to get the best contrast.

At the top right panel I chose Trace Bitmap (another name for an image), then went through the options to see what it could do for me nearest to what I wanted.

Edge Detection gave me an object made of a double line where I chose the thickness between the two, which is an ideal start. Move the original image away. All I have to do now

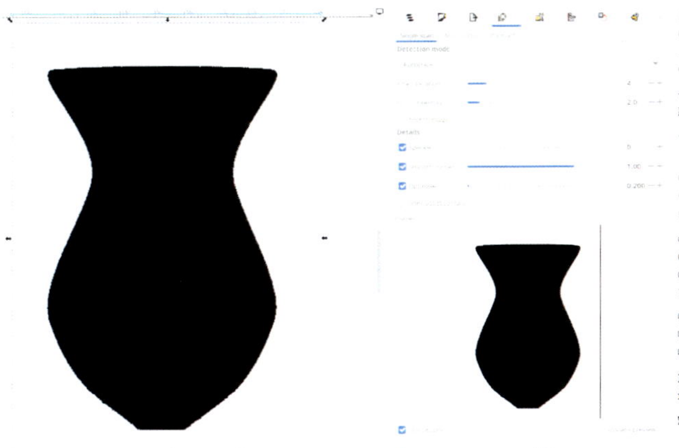

is edit the top using the node tool and I have an outline I can use.

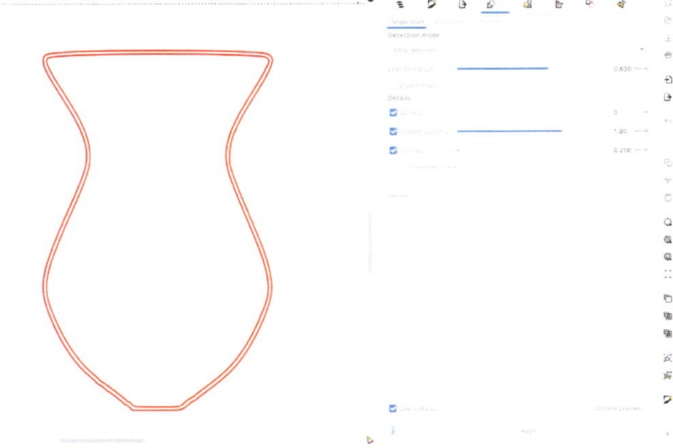

Duplicate

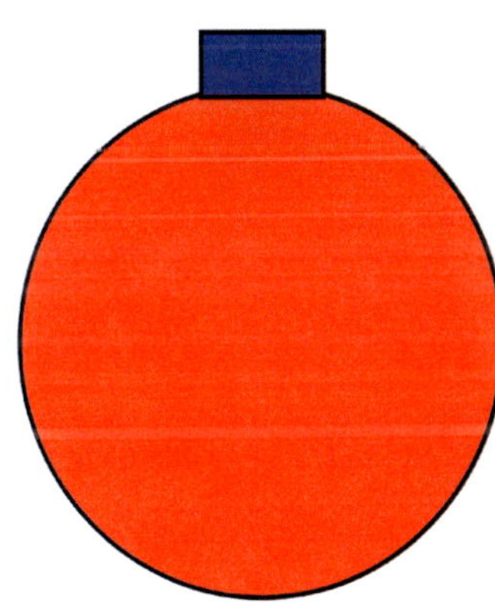

More than just copy and paste, duplicating with a repeated operation lets you create new shapes. As an example I shall make a basic cog by making a circle with a box on top. Also add a tiny circle centred on the first circle. Click on the box twice and a cross should appear - this is the centre of rotation. Drag it to the tiny circle at the centre. Now as you copy and rotate the box it will become a pivot to move the new boxes around the circle.

At the right hand side of the screen click the Transform icon, then rotate and put 45 degrees.

Now Duplicate from the Edit menu and Apply on the rotation box. When you next Duplicate it will keep rotating by 45 Degrees until the circle is full of cog teeth.

It is important to check that things went exactly to plan and touch where you want before the next

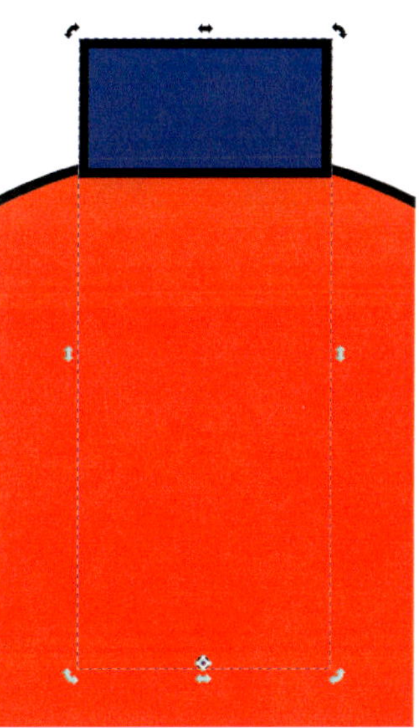

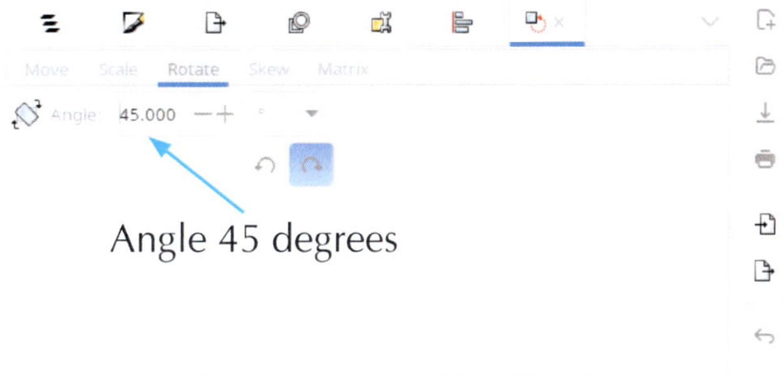

Angle 45 degrees

step so go to View -> Display Mode -> Outline and you will see that the boxes overlap into the circle, adjust them now if not.

Select the circle and teeth and click Path -> Union. You should now see an outline of a cog without the component boxes.

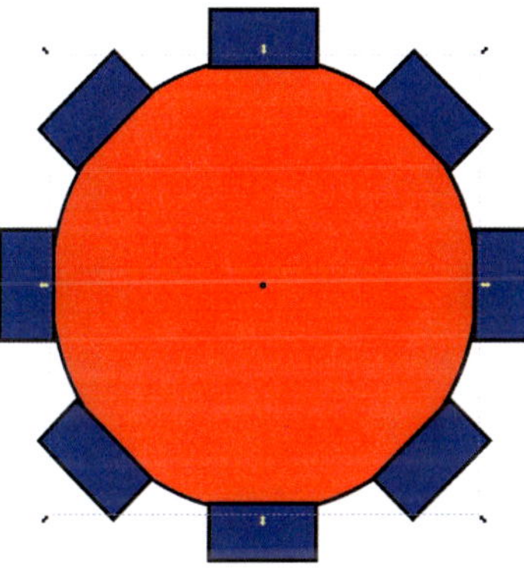

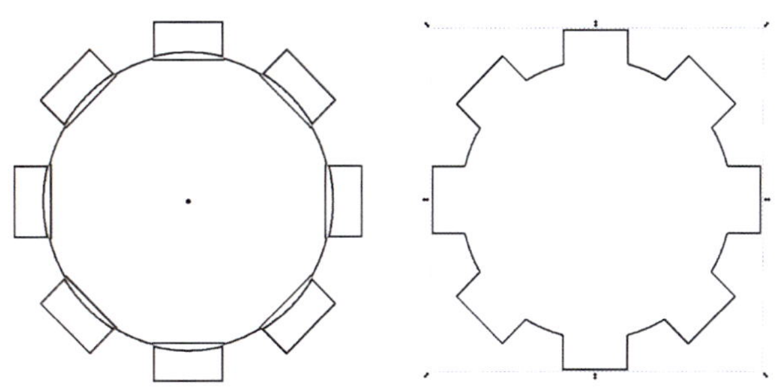

Align

This is available from the Align and Distribute button at the top right. Select two or more objects then chose to have them aligned vertically or horizontally using the buttons.

Distribute

From the same place you can choose to have items equally spaced either horizontally or vertically.

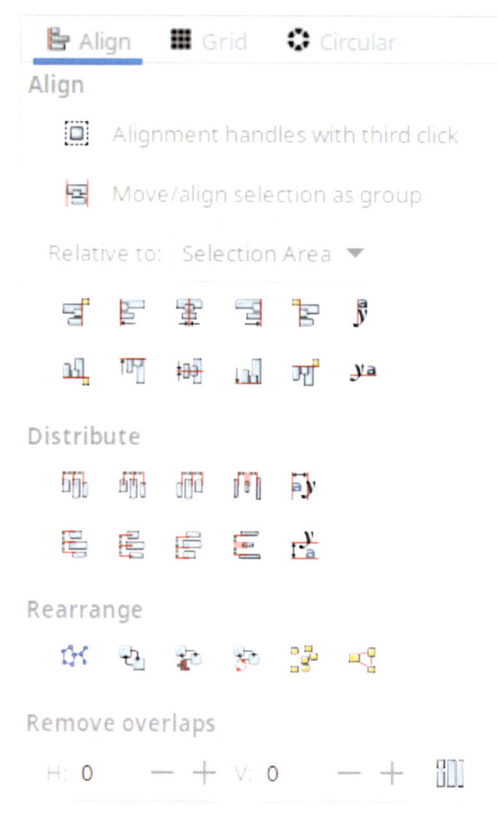

Inkscape is a powerful package available on many desktop platforms and has extensive tutorials available online, here I have just covered the operations most used in this book.

The official tutorials are at **inkscape.org/learn/ tutorials/**

Filler

Filler can be used for perfecting and/or decorating if the surface is not perfect on your print. You can use filler to remove small imperfections. You can also make deliberate imperfections, for example on 'wooden' items and fill with a slightly different hue for a woodgrain effect. Another exciting thing you can do is create spaces for deliberate filling for flooring and other effects. For example for two colour or encaustic tiles etc. Here we've used stucco to create a black and white tile pattern using Inkscape to generate the black pieces. You might have to refill with white a second time because of shrinkage of the filler material, and then sand back carefully with different grit sizes.

Prusaslicer

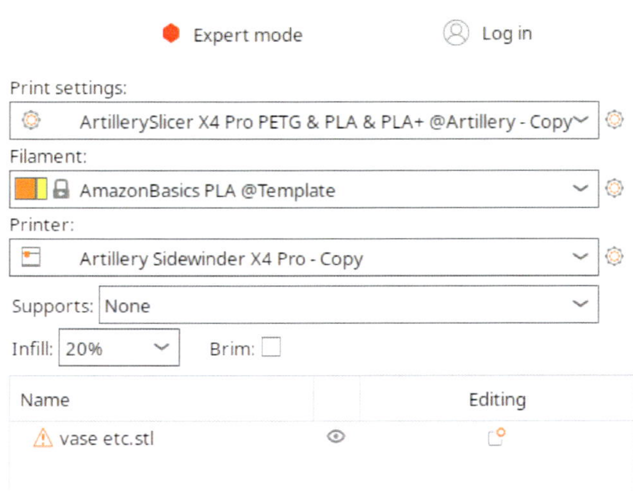

Print settings

You might ask why you need multiple settings as you only have one printer, but for instance the Geeetech comes with lots of presets already set up here e.g. "0.5mm layer height, ultra detail, slow", "0.3mm layer height draft, speed" etc. You can create and save these off for your own printer if they are not supplied

Select material

At the top right you can choose which material you are going to print with. There is a library in the slicer with common filaments that knows their printing and bed temperatures and characteristics.

Select printer

Also there is a dropdown box for printers. The configuration wizard lets you choose the default entry here for new projects.

Add STL to cut mat

Unless you have come straight into the slicer from Tinkercad, the first thing you need to do is put a model on the cut mat. You can do this with the Add button or File -> Import-> STL.

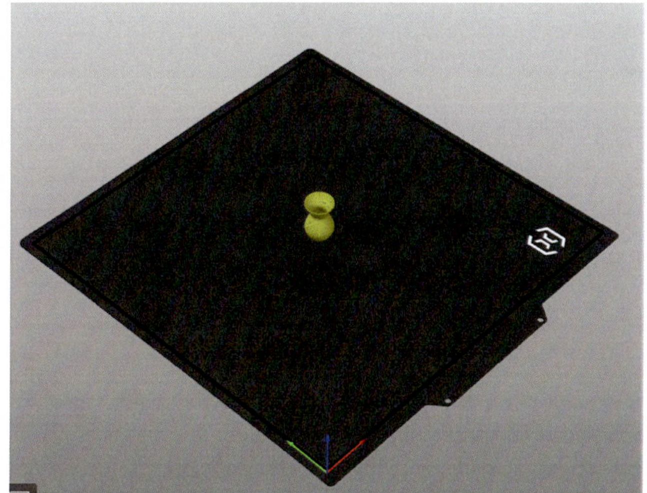

Arrange

When you add more than one item to the mat it might end up with them clashing or even with them all in the same place at the centre. The Arrange button will space them out evenly and sensibly across the space.

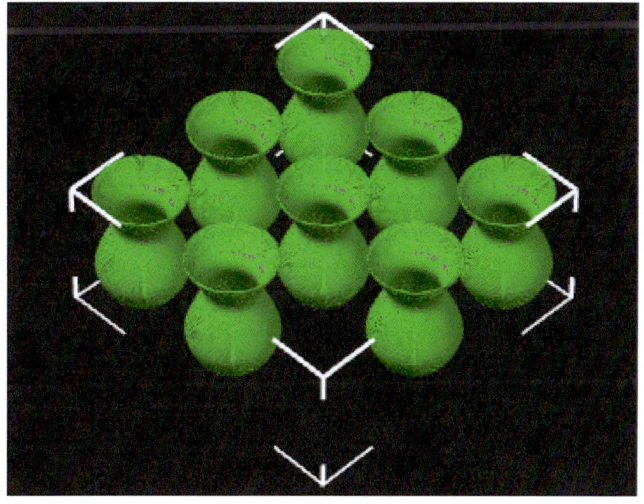

Scale

When I am first working on a project I print all my tests in 24th scale at draft speed. Click on the object, click the Scale button at the left and either manually drag it to approximate size or type 50 into the Scale Factors box at the bottom left.

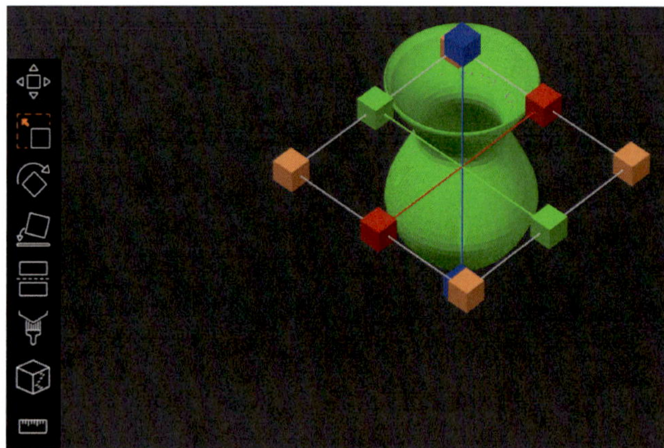

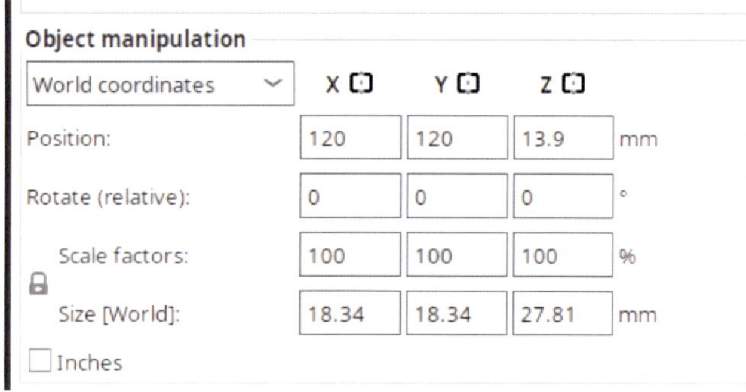

Slice

At the bottom right click Slice Now, it will render up the model to G-code. Then it will provide a panel with estimates of time needed and material usage to give you some idea. These estimates are not perfect but they are a good ballpark figure. The estimates on the printer during printing are often less accurate.

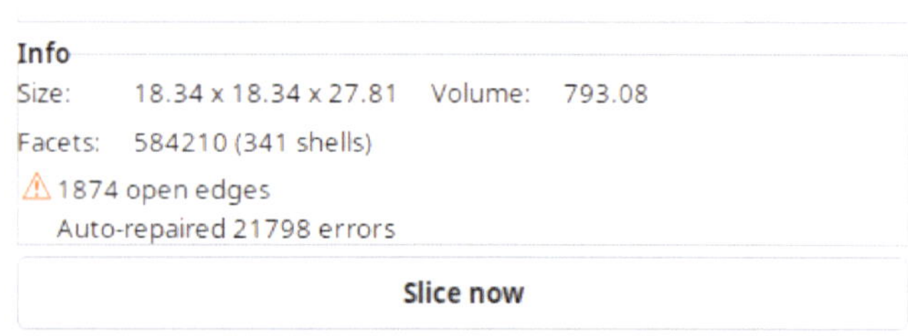

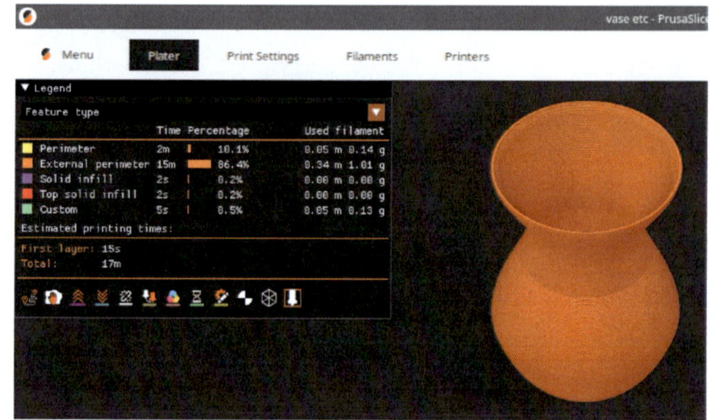

Export G-code

At the bottom right click Export G-code to save off the file. It will be named after the first STL, the type of material, and the estimated time, e.g. vase_PLA_0h17m.gcode

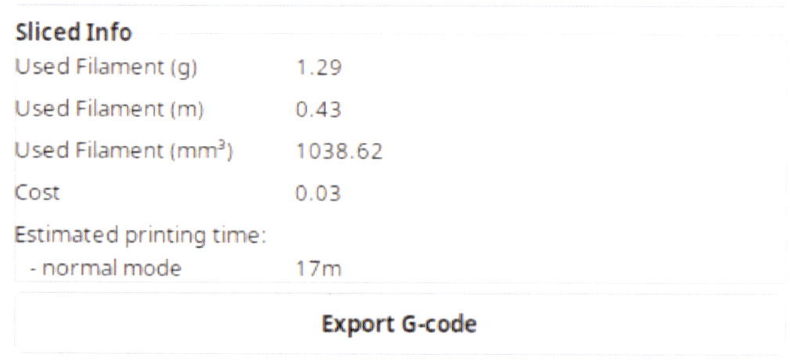

Take to printer

When you have saved the g-code file, either copy it to USB or SD card, or copy it to the printer via Wifi, whichever method your printer uses.

Viewer

If you have badly named some files and are not sure what they will produce, Prusa GCode viewer will show you the contents. This is handy on basic printers with no display panel.

The official tutorials are at
help.prusa3d.com
and click PrusaSlicer documentation

Design basics

If you are intending pieces to interlock, like Lego, make sure the hole is very slightly larger than the piece going into it. I add 0.25mm on the Geeetech.

Pieces that interlock need integral strength, something only 1mm thick is more likely to break than not. And if you print it 5mm thick but have an infill of 20% (the design is filled with 80% air in a honeycomb pattern) then it is still likely to break. I tend to go for at least 50% on these parts. Also the larger you can get away with and conceal, the stronger the design will be. Smaller pieces break if the filament is brittle, especially PLA silk effect.

If you want pieces to print separately but fit inside other pieces then always leave a small margin (e.g. 0.2mm on all sides) and do test prints when you get a printer to know what the tolerance is. You may need to sand the edges with some mats as they leave ridges. My Sidewinder does, my Geeetech doesn't.

If you want to be able to wrap a piece around another piece, a bed of about 0.5mm is strong enough to print and lift off the print bed, but flexible enough to bend fully around a larger object. 0.5mm is made of 0.3mm first layer print and a 2nd layer of 0.2mm. You may be able to go finer depending on your machine.

Similarly if you want to print something flat and bend sections upright later you can rotate a block 45 degrees, make it a hole, and put it 0.5mm above the baseline and it will form a perfect bending point.

It is always easier to have a good idea of what you are going to make before you design it, I like to make scale drawings that I can use later in the process. This gives you a quick idea if the design is in proportion and scaled correctly before you get to the print stage.

Bridging

You will often want to design items with holes in them, this is called bridging when you expect the printer to print a solid section over a hole. This file lets you test how far you can go without the bridge breaking apart. You can then tune your settings until you are happy or have to admit the limit is reached.

www.thingiverse.com/thing:476845

3D printing problems

Stringing

This is where there are lots of strings or fibres between uprights in the print, causing a mess you have to cut away. It can be caused by a few things but the first ones to check are filament temperature being too high and retraction.

Temperature: as a rule use the high range for the first layer and the low range for the rest, e.g. if PLA says "190C - 220C" use 220C for the first layer and 190C thereafter.

Retraction: some machine designs need a different setting to others, the ones with a tube may need more e.g. 5mm retraction of the filament during moves.

Blocked Nozzle

I don't poke pins in it or mess around with any other methods. First if it was blocked with PLA I take the heat of the nozzle up higher than

220C and see if I can clear it out that way. If not I take the nozzle off. The all metal ones are fine in a bit of tin foil in a very hot oven for 3-5 mins, that will melt out any filament left in there perfectly.

Imperfections

If the surface is not perfect on your print you can use filler to remove small imperfections You can also make deliberate imperfections, for example on 'wooden' items and fill with a slightly different hue for a woodgrain effect. Another exciting thing you can do is create spaces for deliberate filling for flooring and other effects. For example for two colour or encaustic tiles etc Here we've used stucco to create a black and white tile pattern. You might have to re-fill a second time because of shrinkage of the filler material, and then sand back carefully with different grit sizes. The possibilities for your own designs are endless.

PLA+

This labelling caused me awful problems first time I loaded it. The + was tiny and I didn't notice it. It loaded OK as PLA (my printer has presets for preheating the head). It even extruded the first layer. Then after that it locked up because it needs a higher low end temperature and a lower high end temperature. So I unblocked the nozzle as above. Still no movement. I found that while not extruding the small wheel feeding filament in had been wearing away at the filament until there was a groove in it and it no longer moved in or out. Now I preheat to the right temperature.

Change Filament

This is hidden on my Geeetech printer unless you have already preheated. Which is a nuisance isn't it? Why not always be there and preheat when it is needed like on the Sidewinder? It preheats during the change over, why not at the beginning too? I suppose that would depend on the type of filament but I find hiding important options just leaves people struggling and pulling it out by hand. And if the head isn't heated when you do it that can cause damage.

MINTEMP error - bed or head

All those moving parts is bad news for cables which prefer to be static. The constant movement can easily cause a cable to break after a while, the ones to the temperature probes are very fine wires (often 22 AWg / 0.3mm2) and particularly prone to it.

See pictures for examples of a bed error I got in the misadventures section later, the culprit was a lead broken in two right where the bed moves against the cable loom. The thin wires come from the temperature sensors, the thick ones heat the bed.

Threading a repair cable through the sheath for the thin red wire and splicing it to the current one at both ends seemed to do the trick.

Runaway Thermal error

So after fixing the Mintemp error a little while later I got a different error but with similar symptoms to the Mintemp error - it would occur when the bed reached a certain position. So I disassembled it and after careful testing found the thick red wire did not conduct in certain positions. Another wire needed to be spliced in.

And a couple of weeks later the same thing again - the cables weren't secured *savagely* tight with cable ties, just tight. All 3 of them had broken just after the cable tie too, where the cable is bent on full bed retract. So I replaced the heating wires which were 1.2mm or so with 2.5mm cable and taped before and after the cable tie to strengthen it a little. Fingers crossed.

Display corruption

After these repairs I found the display started turning to gibberish on some prints. Seems to be the shielding on the ribbon cable can just rub off after a bit of handling (like taking it on and off to get in to repair the cables). I have lightly wrapped it with aluminium tape.

Bed temperature keeps rising

This might be caused by a broken wire or broken MOSFET. You will likely need to search support forums for the problem on your model of printer to find out the most likely path to fixing it. A multimeter is cheap and essential to diagnosis here. You can check if a wire is broken with it, or if a MOSFET is fixed to "on".

Warping or prints flake off the bed during print

Warping is where the corners of a print bend up during printing, or even lift off losing the print entirely. The very first thing you should always check with this kind of problem is if the model is at zero on the workspace! Then check bed levelling. If the level is off the filament will be laid in mid air, not to the bed.

Next check everything is still tightly screwed together. My Artillery Sidewinder goes so fast on infill that it started to shake its arm strut loose! A quick tightening and everything was good again.

Then there is also bad adhesion which can have a number of causes and fixes. First clean the bed: scrape off any residue, then clean with alcohol. Try it again.

If this doesn't work raise the bed temperature a little, say 5C at a time. If it still flakes off use a sticky coating such as the readily available glue sticks.

If all else fails add brim in the slicer settings. This is fine for larger prints but can be a nuisance to clean off perfectly on miniatures.

Print sticking to bed too firmly

If the print (resin or filament) seems too firmly attached to securely remove it (you might break the print or excessive force may damage the printer or bed) then normally you can just put it in the freezer for a while. The change in temperatures is handled differently by the two materials - the metal shrinks faster, and this helps loosen the grip.

Disjointed prints

If part of the printer has come lose it may cause the print to lose its positioning. I found the Sidewinder moves so fast the fixings loosen every few weeks. This has caused prints to fail as the machine wobbles and misses the position it is aiming to get to.

My problem appears to be the "clever" part of the Sidewinder that automatically checks the bed level had gone very wrong. When I tried to level the bed after checking the screws were tight I couldn't make free space to drag paper between bed and nozzle. I had to star from scratch and tune the Z Offset, then manually level, then auto level. Then ... oh just see the misdaventures section for the picture of smoke.

Cheap electronic components

The control board of cheaper printers can be shipped with the absolute cheapest components that control the power to the heating elements (MOSFETs). Getting that few-cents part replaced with one costing a few cents more takes time, postage, and tinkering with a soldering iron.

First you have to find the MOSFET in question so you will have to find a board diagram. It can be covered by a heatsink too, which makes it easy to spot but can be harder to get access from the top side. Once identified, desolder it. This can be easiest with desoldering braid dabbed repeatedly over the pins and wicking away the solder. Carefully remove the existing

MOSFET (if it isn't pulling easily do more desoldering) then solder the new one in and trim its pins to length. At this point if you make an awful blobby mess with solder like me, a dab or two of the braid will make it look like a pro did it.

Factor this into your price calculations when buying, yes it may save you a hundred dollars/euros/pounds but you may well pay that out in delays and frustration later.

Choosing a printer

There are two main types: filament and resin. Resin tends to have a much smaller printing bed but be great at fine detail. Filament can have huge beds capable of printing large objects (for a miniaturist!).

Whichever road you take, make sure to see reviews of the one you choose first. And also poke around the support forums with your printer name followed by "problem" to see what you might be likely to face a few months down the line.

Is it reliable?

My first one (Geeetech A10) ended up in bits to be repaired more often than working. Yeah it was easy and cheap to get the parts but still a pain and not for the faint hearted. My next was an Artillery Sidewinder X4. It's mid-priced and very fast, but if money was no object I would have gone for Bambu Labs or Prusa, both with a reputation for precision and reliability but both at the high end of the price range.

Miniaturists using resin seem to choose Elegoo, the only question being about which size and which addons (e.g. washing stations). I went for the Mars which is middle sized.

Size

What are you looking to make? Tiny pots, pans and detailed miniatures are fine on any size of bed such as the smaller resin printers, but if you want to make furniture and dioramas you will likely want a filament printer with a bed of at least 20cm/8" square.

Try to be sure you are heading down the right path if you only intend to buy one type.

What software does it use?

Does common filament slicer software support it (e.g. Cura, Prusaslicer)? Otherwise you may have to get technical and build your own settings profile (all the pages of options to make a slicer work with your printer to its best). Expert users recommend this of course as you can tune the best performance and results from a machine, but I prefer to be able to print *something* straight out of the box just to know I am on the right track.

The Elegoo resin printers come with ChituBox though it does support others. This seems to be more than capable but I would prefer to only be using one slicer in my life!

How do I control it?

Do you have to connect a computer? If so, what operating system and version? Can I use a USB stick or card?

I prefer the standalone options so they aren't tied to often-crashing computers to work. My Geeetech, Artillery and Elegoo all allow USB or SD card working. The Artillery also allows wifi sending and printing of files.

The Artillery and Elegoo also have a nice colour screen to see what the files will print before committing which is handy if you have a lot of parts of a project on one device.

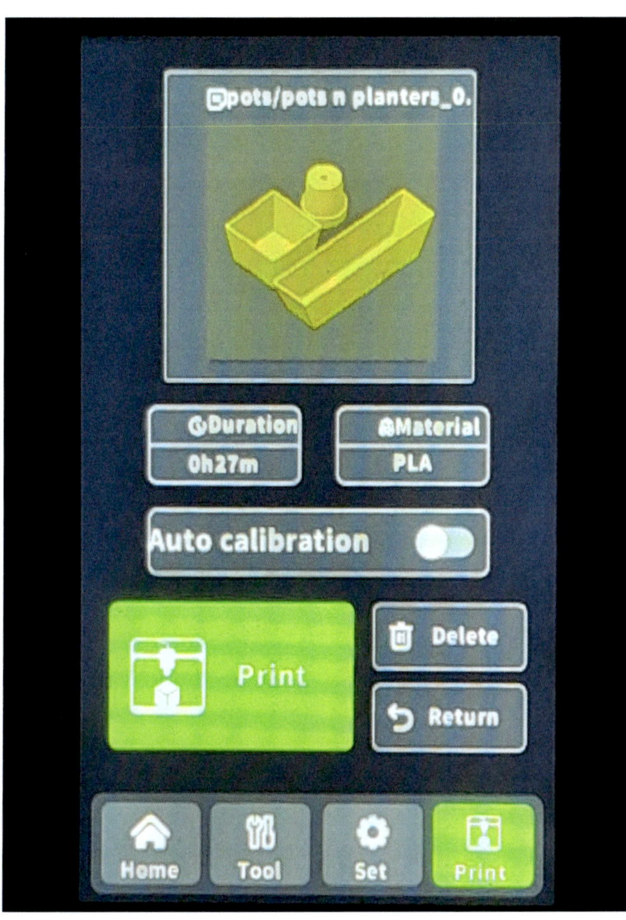

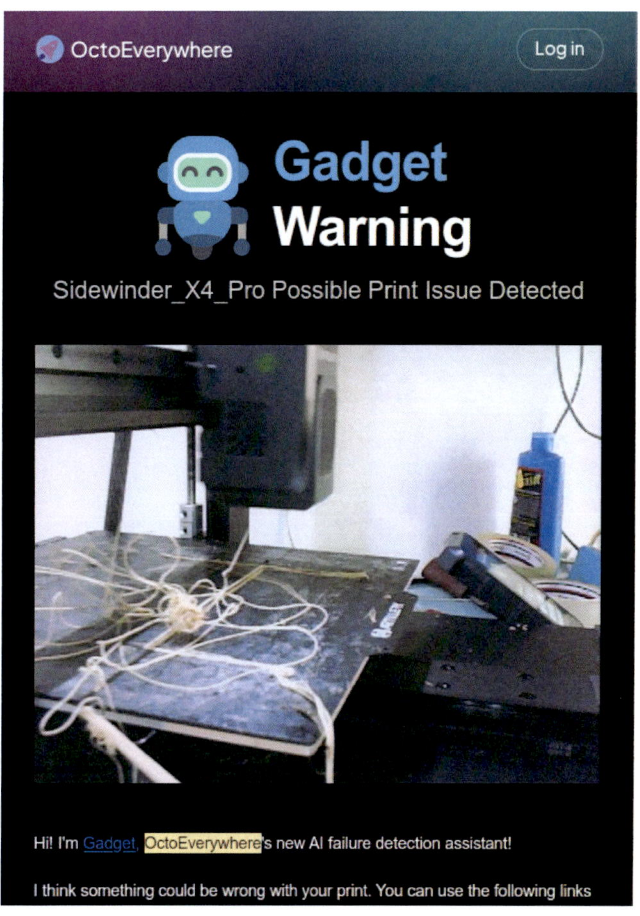

Monitoring prints

Can you monitor and control it remotely with common software, like Octoeverywhere (free trial, then free to monitor progress, monthly subscription for video streams)? This lets you watch your print from a phone or computer using a webcam and see all the progress info too. The Artillery runs on Klipper (Linux) and thus lets you install these features and others. You can get email or app alerts when a long print has finished etc.

What materials does it support?

Some clever filaments are available that let you do complex prints and wash away parts that support "mid air" pieces, letting you print a piece all in one. Is it something you want to do? Do you want to use high temperature filaments for strength or flexibility? Check you have the option.

What mat does it have?

Seems an odd question, but my Geeetech came with a perfect, smooth, fixed mat. This can cause issues with slipping if you aren't properly levelled and set up but on the whole it is the best choice for miniatures.

My Sidewinder came with a textured mat. It was magnetic (you can pop it off and flex for stuck prints) and double sided too which is great, however it leaves a "crackle effect" across everything and so smudges the edges a little at the mat. This has been a problem when making tight fitting components, and led me to trawl the internet for hours to find a smooth mat. Here I have a double sided mat from Temu that was very cheap, here's the smooth side - remove the film, clean with alcohol and re-level the bed.

Patron questions and answers

We asked Angie's Patreon subscribers, many of whom are makers themselves, if they had questions (and answers!)

Q1: How much do they cost?

A: Not much for entry level machines but it can range from $£200 for a cheap Geeetech / Creality to $£1000 for the big names.

Q2: Where do you get them from?

A: Amazon for cheap, specialist sites for better machines.

Q3: Do they break very often?

A: Yes they break, sometimes in many different ways, anywhere cables move can be a problem, anything with cheap components too.

Q4: If they do break, is it fairly straightforward to maintain/fix them, or do you need to hire someone to do it?

A: On filament ones all the parts are modules and cheap to replace, though some cables or parts may require a bit of soldering.

Resin printers have very few moving parts, the screen would be the worst thing to damage because of the expense.

Q5: Is the deal the same as 'real' printers - i.e. the machine itself seems like surprisingly good value... until you realise how much printer cartridges cost?

A: Amazon basics filament is cheap and goes a long way, others like Anycubic seem good too. Resin can be expensive but it makes a lot of miniatures.

Q6: If I wanted to buy one, what should I look for (or avoid) in a machine?

A: The brand of cheaper machines is just a personal choice - Geeetech and Creality seem to be mostly made from the same commodity parts. After that midrange or high end is down to your price point.

Q7: Do you have to buy special software to create the designs, or can you find free/cheap software?

A: All the software you need is free, though you can pay more if you find something that works better for your way of thinking.

Q8: Can you split the STL files for people who have smaller beds on their printers?

A: Yes an average filament printer should be 20cm/8″ square but resin is normally much smaller, so there is even free software to split it, and not so free that is used for really big prints.

You can also quickly chop files up in Tinkercad using blocks marked as holes - they make whatever they cover disappear from the print. So 3 blocks covering a boat lets you make 3 sections of 10cm/4″. Just group to make pieces disappear, export, ungroup and disappear a different section. If you are practised this is often quicker than opening new software. See the example in the shop sign section earlier.

Q9: Resin printers owners: do you have the air purifier attachments?

A: (a) I have those little towers with carbon filters that you put inside the cover while you print. They do a good job with the smell, but I don't know if they work with anything odourless

(b) I have mine in the garage, without an air purifier or any extraction system. I do put on a mask, in addition to gloves, and glasses when I use it.

Q10: Resin printer owners - do you have an all in one washing and curing station?

A: I have the Elegoo all in one - it was the best thing I bought - so easy to post-process the prints.

Q11: I tried to learn Shapr3d / Blender but I found it quite hard, and the only design program I truly know is Tinkercad, which is a bit basic for the things I'd like to be able to design.

A: Tinkercad is great for assembly but I use 2D software to make SVGs of the complex parts where I can. It has some addons tucked away that let you do a lot of the clever things from other software, the main thing I had been missing was a "lathe" effect letting you turn a 2D line into a 3D shape such as a table leg. I found an addon called SVG Revolver (tutorial https://www.tinkercad.com/things/jbGs7BTCBCw-) which should be available in Tinkercad by default. However my methods earlier for the vase can be easier to get results.

Q12: What's the strongest filament?

A: I found any "silk" PLA to be awfully weak and brittle, which is a shame as it looks shiny and I did some big prints in it for a drum carder. Plain PLA is better though still a little brittle, PETG is probably the best choice. Other materials are available but not supported by low end printers.

Other Q&As supplied by Tommy Hogg

Tommy is a long time 3D printer user who is not afraid of a soldering iron.

How do I get started with 3D printing?

To get started, choose a beginner-friendly printer, learn to use slicing software, find or create a 3D model, and start printing!

Where can I find 3D models?

3D models can be found on websites like Thingiverse, MyMiniFactory, and Cults3D, among others.

How long does 3D printing take?

Print times vary widely based on the object's size, complexity, and printer speed, ranging from minutes to several days.

What is the best 3D printer for beginners?

Popular beginner-friendly printers include the Creality Ender 3, Prusa Mini, and Anycubic i3 Mega.

How do I store 3D printing filament?

Store filament in a cool, dry place, ideally in airtight containers with desiccants to prevent moisture absorption.

What causes print failures?

Print failures can result from bed adhesion issues, incorrect temperature settings, clogged nozzles, and poor model design.

How do I maintain a 3D printer?

Regular maintenance includes cleaning the nozzle, checking the bed level, lubricating moving parts, and ensuring the extruder is functioning properly.

What is a cooling fan?

A cooling fan helps to cool the filament after extrusion, improving layer adhesion and reducing stringing.

Can I print in color?

Yes, color can be achieved by using multi-material printers, changing filament during printing, or painting the finished model.

What are 3D printing services?

3D printing services allow you to upload your design and have it printed and shipped to you, useful if you don't own a printer.

What is post-processing?

Post-processing involves finishing techniques like sanding, painting, or assembling parts after printing to improve aesthetics and function.

What safety precautions should I take?

Always operate printers in a well-ventilated area, be cautious of hot surfaces, and use appropriate protective gear when handling resin.

How do I clean my 3D prints?

Cleaning depends on the material; FDM prints can be washed with warm soapy water, while resin prints require isopropyl alcohol.

What is the difference between open and closed frame printers?

Open frame printers lack an enclosure, allowing better airflow but less temperature control, while closed frame printers maintain a stable environment.

What is 3D scanning?

3D scanning captures the shape of a physical object to create a digital 3D model, useful for replication or modification.

What is a dual-extrusion printer?

A dual-extrusion printer has two nozzles, allowing it to print with two different materials or colors simultaneously.

How can I improve print quality?

Improving print quality involves fine-tuning settings, using high-quality filament, ensuring proper bed leveling, and maintaining the printer.

What is a filament sensor?

A filament sensor detects when filament is running low or has run out, pausing the print to prevent failures.

Misadventures with a cheap filament printer

The problems section earlier hinted at the chequered history of my Geeetech A10 repairs - here are some pics. First the butchered mess of the heating and temperature probes after 3 repairs or so. Each time grafting new pieces of wire and finally replacing the heating wires with much heavier duty ones.

Next the MOSFET problem, a picture of me attempting to remove a heatsink soldered right above the MOSFET making access very difficicult from above to replace it, then a picture of the new MOSFET before it went in.

The final Geeetech picture is the state of the ribbon cable connecting the display Before I

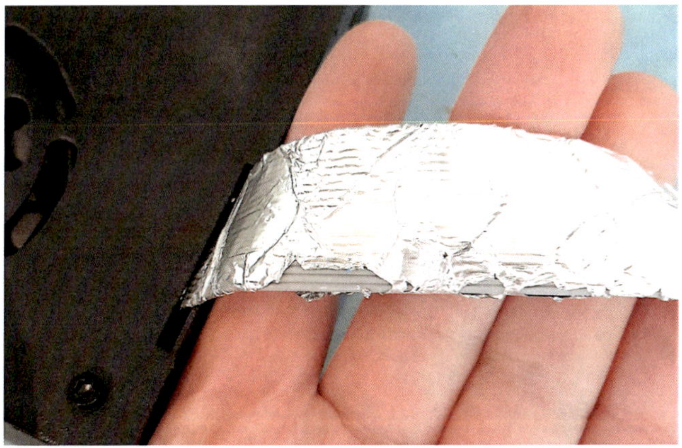

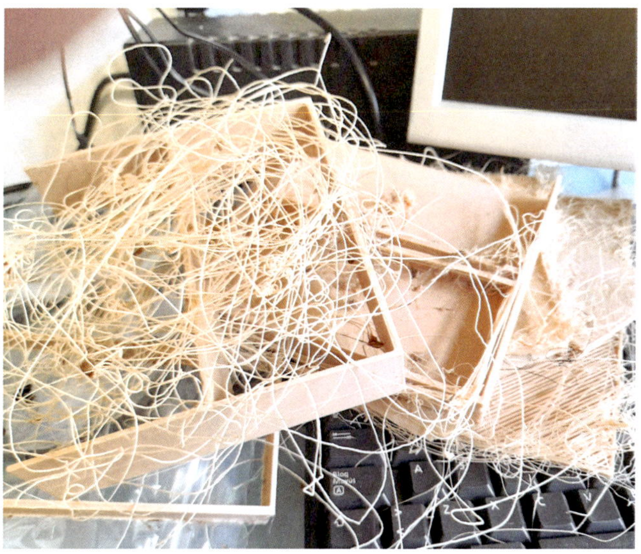

replaced the aluminium tape. This tape deflects interference and the display gets garbled without it.

And just before the last prints of the book the Sidewinder started producing spaghetti instead of prints ... and smoking! I am not sure this photo does it justice, the video is more dramatic. Anyway smoke started coming from the head. Support say this is a likely failure of the mainboard and the heating end.

For this reason I am loathe to leave the house with a print going. If it is just a failure I can press stop, but a runaway thermal event in the head can be very serious.

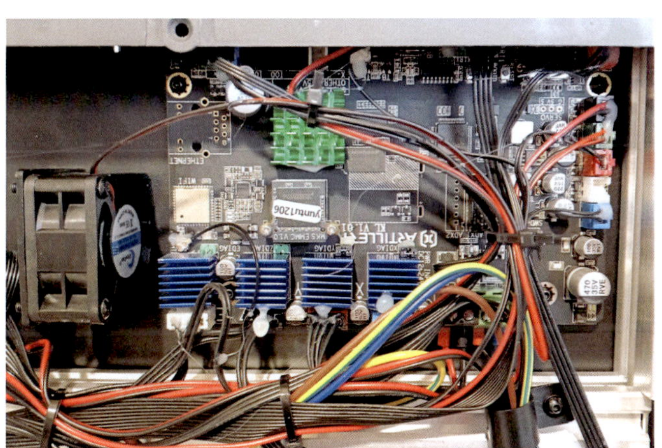

Anyway one free replacement motherboard, one free replacement hotend, a firmware update and I appear to be printing again as good as new.

Glossary

3D printer   the machine making the 3D objects from resin or filament

3MF   the standard format for laying out 3D models for slicing - 3D Manufacturing Format

ABS   an older type of filament with fumes issues - Acrylonitrile Butadiene Styrene

Brim   a small detachable layer around an object to make its footprint larger for stability during the print.

Bed   the flat area that a filament printer deposits material onto as the head is moved away vertically creating the layers, this bed is normally heated for better adhesion.

Build plate   this is the equivalent of a bed for a resin printer where the material is accumulated, it normally hangs upside down and is pulled away from the screen.

FDM   the process of printing using filament - Fused Deposition Modelling

Filament   the plastic material on a reel

G-code   the instruction set used by printers to tell them where to print, how fast, how hot etc.

Infill   how much filament to use in large blocks of print

Layer height both types of printers work in a similar way, building up the print layer by horizontal layer. You can choose the layer height for speed or precision.

Nozzle   the small metal end of the extruder that restricts the size of plastic that can leave, a normal default nozzle is 0.4mm

PETG   a filament that is slightly stronger than PLA - *PolyEthylene Terephthalate Glycol*

PLA   a common 3D filament without so many fumes during use - *PolyLactic Acid*

Resin   photo sensitive liquid material

SLA   the process of printing using resin - *Stereolithography*

Slicer   the software that takes 3D models and converts them to instructions to extrude filament

STL   the standard for 3D models - *STereoLithography file format*

Stringing   this is a problem where fine traces of filament build up between points of the print in a fine web

Supports   if you need to print a design that hangs over into open space you may need to use supports, which are lightweight pieces of print that should be easy to break away from larger designs

SVG   the standard for precision 2D illustrations - *Scalable Vector Graphic*

TPU   a strong filament that can take impacts - *Thermoplastic PolyUrethane*

Warping   this is where part of a print lifts off the bed during printing

Z Offset   part of levelling the bed this is setting the height of the nozzle from the surface as the default 0 height.

Biography

Frank Fisher is best described as a 'creative technician'. Since the very early days of the internet he has always helped musicians, artists and writers do whatever it took to publish or reproduce their work. Producing websites, music, videos and books, along the way doing all the IT work it took to turn those creatives dreams into physical and saleable forms. Frank fills in the holes in the IT knowledge of his clients and often his friends and family. His love of and ability in 3D modelling came from doing 3D animation in the 1990s when the hardware was very, very slow! Turning multi dimensional ideas into instructions for information and printing technology is what he does best and what he enjoys.

Frank lives an alternative off grid-in-a-village lifestyle. Alternative because there is no 9-5, and it pairs up-to-date technology with frugal-living. Not always an easy balance.

Frank lives and co-works with his multi-creative co-author and wife Angie who is a well known (in the small world of miniatures) craft and craft business writer. They share the home of too many cats.

Thanks and acknowledgements

Tommy Hogg for constant support and encouragement of many kinds during my printing experiments

The patrons without whom the past few years would have been much more difficult if not impossible and none of the last few books would have been written. You are all our supporting angels!

Lisa Sones-Peck, Christine McKechnie, Jacquie Hall, Maggie Dokic, Rachel Timmerman, Natalie Bartin, Robyn Stewart and, of course, many more who didn't give permission to be named but who are just as important to me. Thank you all!

Patreon
Why we love our patrons

If you've never heard of Patreon before, imagine you could be a 'patron of the arts' in some small way helping your favourite artists to continue working, inventing and teaching in their specialist area. Artists no matter how well known in their field often have no regular guaranteed income and often give away their inspiration for free because until now there wasn't an easy method to gain an income from day to day teaching, support and skill sharing.

This subscription service is an easy way to connect artist teachers with their students and followers, and as a way for the 'patrons' to give the level of support they are comfortable with and receive in return (sometimes personalised) perks such as early access to new ideas, live patron only videos and little samples of work to help you visualise stages of work and qualities of colour. As well as advance knowledge of really new ideas before they ever get to publication. Some ideas of how I made things which never even reach the books which I call my 'daft ideas' for example how I made the awning for the dollhouse shop on the front just using parts from an old umbrella! For benefit patrons I'm also able to send out little found 'things' which might inspire you, or samples of my new tools before they go into full production.

Many thanks to my current 70+ patrons some of who have been with me for several years now. You've all given me courage to start with new things like this book. The Patreon thing has really helped me because it's like having 70 sets of shoulders to lean on. 70 therapists and 70 special friends to share my daft ideas with and see if they work. Or at least are interesting enough for you not to walk away! 70 people who understand that no matter how well known an artist is they still may struggle from time to time. That's worth so much!

www.patreon.com/angie_scarr

Making Miniature Food & Market Stalls

Angie's first book Published by Guild of Master Craftsman Publications. A bestselling introduction to making polymer clay miniature food. This is an updated edition.

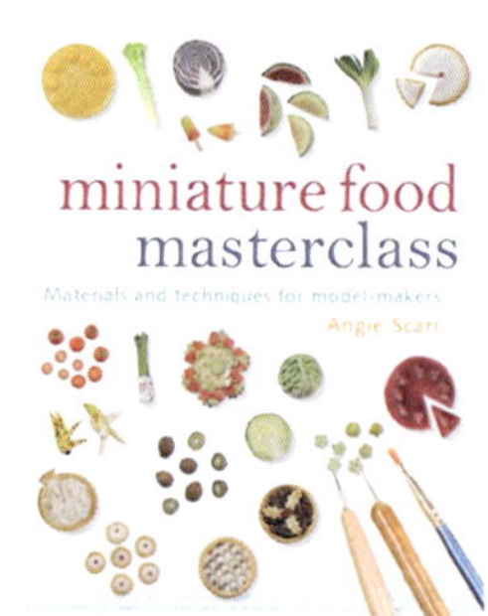

Miniature Food Masterclass

Angie's second book with GMC. Also still a bestseller this one continues the journey of exploration into what polymer clay can replicate.

Other books by Sliding Scale

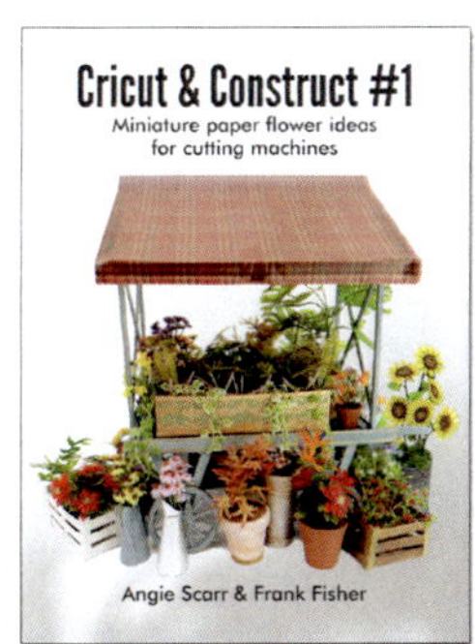

Cricut & Construct #1

How to design SVGs and cut paper flowers and plants on Cricut, Silhouette, laser etc.

All the finished designs are available on our Etsy store.

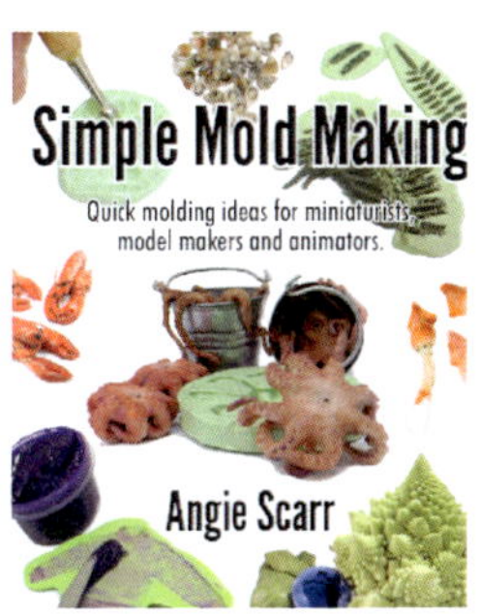

Simple Mold Making

A book full of quick molding ideas for miniaturists, model makers, animators and jewellers using 2 part (Silicone) mold material.

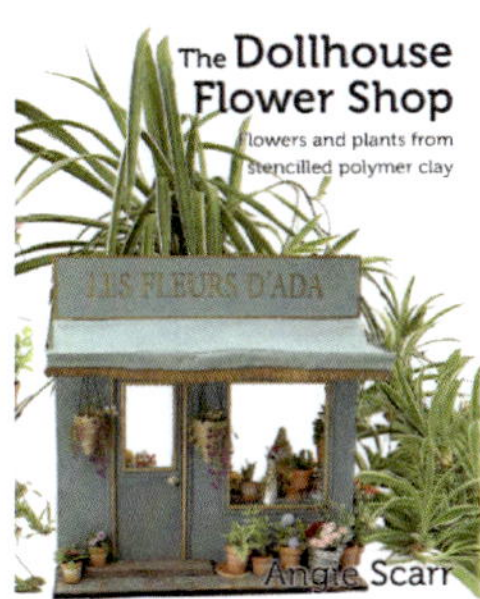

The Dollhouse Flower Shop

This book concentrates on the innovative idea of stencilling flowers in polymer clay/liquid clay mix. Some equipment and materials are needed to get started.

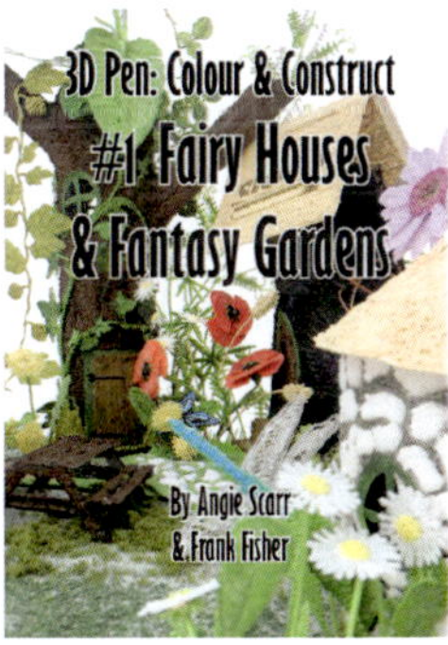

3D Pen #1 Fairy Houses and Fantasy Gardens

A handy pattern book for anyone of any age who is looking for a project to make with their 3D pen. Excellent addition to a 3D pen gift.

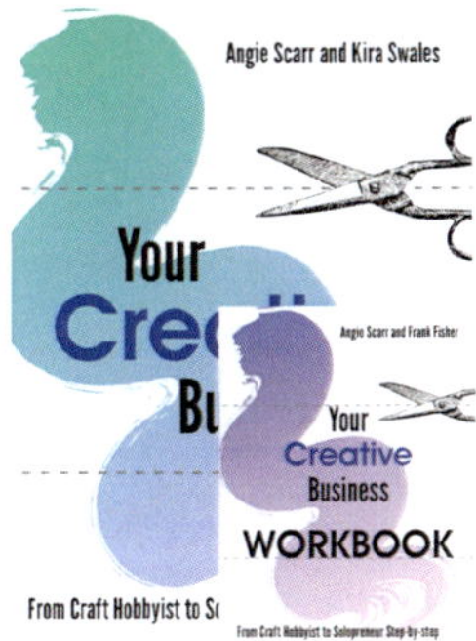

Your Creative Business Books

Angie and SEO expert Kira share advice on all aspects of craft business from pricing and marketing through to multiple income streams to help you ensure a more secure future.

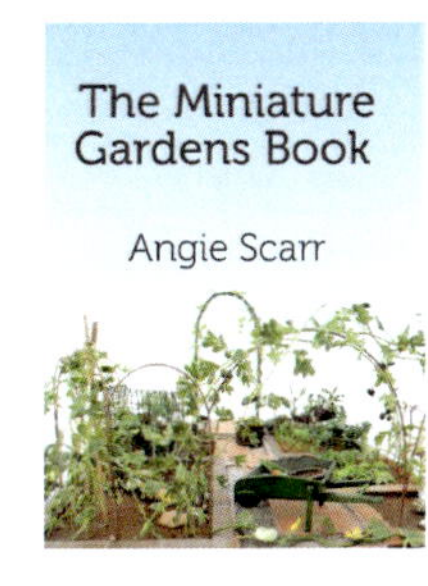

The Miniature Gardens Book

Have you ever fancied making more than just a flower garden in miniature? Angie gives you several garden styles and lots of new ideas.

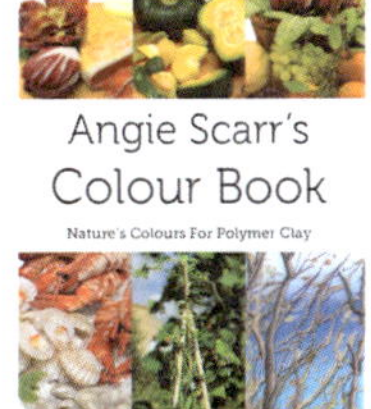

Angie Scarr's Colour Book

New large edition book that asks the big questions about colour realism in polymer clay, helping you towards work that is so realistic it jumps out from the rest.

Angie Scarr Miniature Challenges Parts 1 & 2

Revisiting all the old magazine articles in Dolls House and Miniature Scene and other dollhouse magazines. Some with updated information.

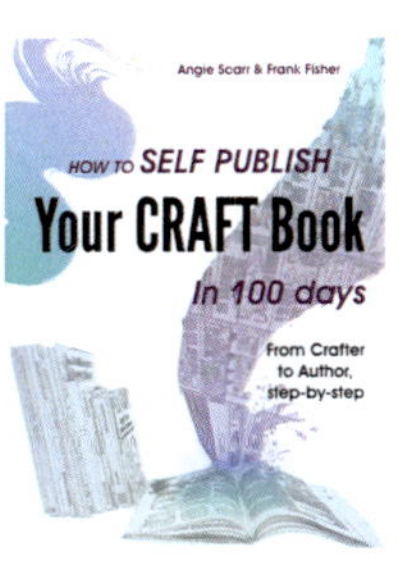

How To Self Publih Your Craft Book in 100 days.

A step-by-step guide through all the stages.

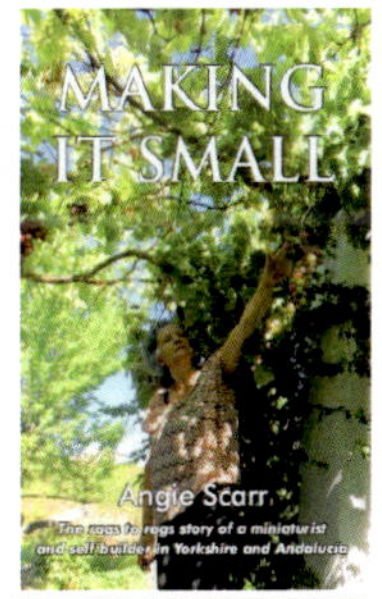

Making it Small- Biography

Angie never lived an 'ordinary life'. When she and Frank met it became less ordinary still. A story of the love of crafts, miniatures, self building and life in a small pueblo in Spain.

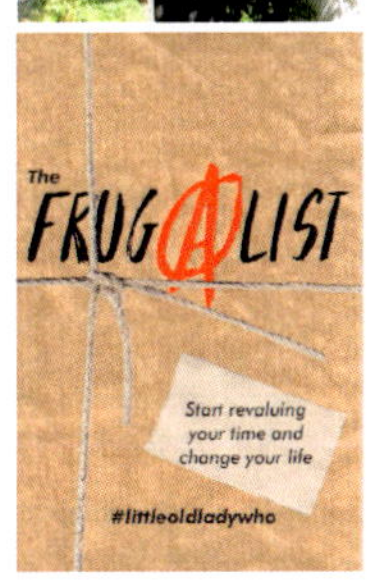

The Frugalist

A look at revaluing your time, living better for less and gently preparing for unexpected crises. If life sometimes feels tough this book might just help.

 patreon.com/angie_scarr
Support my writing

 pinterest.com/angiescarr
Links to my work all over the internet

 facebook.com/angiescarr.miniatures
My facebook page where I let everyone know what is going on

 ko-fi.com/angiescarr
If you've learned something valuable buy me a coffee

 instagram.com/angiescarr
Photos of work in progress

 tiktok.com/@angiescarr
Video shorts

 youtube.com/user/angiescarr
For tutorials, howtos and videos about crafts and miniatures

 @littleoldladywho1.bsky.social
Bluesky: the new Twitter

 **etsy.com/shop/AngieScarrCrafts**
Our Etsy digital store for plotter / cutter files / 3D printer
Spiral staircases, market carts, flowers, leaves, boxes, flowershops and more

 **angiescarr.com**
For moulds, stencils, kits, books, miniatures and other craft materials
Delivery worldwide and trade terms available